S0-BBO-734

PROSPERITY'S PROMISE

Dellplain Latin American Studies

No. 12 *Credit and Socioeconomic Change in Colonial Mexico: Loans and Mortgages in Guadalajara, 1720–1820*, Linda Greenow

No. 13 *Once Beneath the Forest: Prehistoric Terracing in the Río Bec Region of the Maya Lowlands*, B. L. Turner, II

No. 14 *Marriage and Fertility in Chile: Demographic Turning Points in the Petorca Valley, 1840–1976*, Robert McCaa

No. 15 *The Spatial Organization of New Land Settlement in Latin America*, Jacob O. Maos

No. 16 *The Anglo-Argentine Connection, 1900–1939*, Roger Gravil

No. 17 *Costa Rica: A Geographical Interpretation in Historical Perspective*, Carolyn Hall

No. 18 *Household Economy and Urban Development: São Paulo, 1765–1836*, Elizabeth Anne Kuznesof

No. 19 *Irrigation in the Bajío Region of Colonial Mexico*, Michael E. Murphy

No. 20 *The Cost of Conquest: Indian Decline in Honduras Under Spanish Rule*, Linda Newson

No. 21 *Petty Capitalism in Spanish America: The Pulperos of Puebla, Mexico City, Caracas, and Buenos Aires*, Jay Kinsbruner

No. 22 *British Merchants and Chilean Development, 1851–1886*, John Mayo

No. 23 *Hispanic Lands and Peoples: Selected Writings of James J. Parsons*, edited by William M. Denevan

No. 24 *Migrants in the Mexican North: Mobility, Economy, and Society in a Colonial World*, Michael M. Swann

No. 25 *Puebla de los Angeles: Industry and Society in a Mexican City, 1700–1850*, Guy P. C. Thomson

No. 26 *Generations of Settlers: Rural Households and Markets on the Costa Rican Frontier, 1850–1935*, Mario Samper

No. 27 *Andean Ecology: Adaptive Dynamics in Ecuador*, Gregory Knapp

No. 28 *Disease and Death in Early Colonial Mexico: Simulating Amerindian Depopulation*, Thomas Whitmore

No. 29 *Veracruz Merchants, 1770–1829: A Mercantile Elite in Late Bourbon and Early Independent Mexico*, Jackie R. Booker

No. 30 *Laricollaguas: Ecology, Economy, and Demography in a Seventeenth-Century Peruvian Village*, David J. Robinson

No. 31 *Encomienda Politics in Early Colonial Guatemala, 1524–1544: Dividing the Spoils*, Wendy Kramer

No. 32 *The People of Quito, 1690–1810: Change and Unrest in the Underclass*, Martin Minchom

No. 33 *Demography and Empire: A Guide to the Population History of Spanish Central America, 1500–1821*, W. George Lovell and Christopher H. Lutz

Dellplain Latin American Studies

PROSPERITY'S PROMISE

THE AMAZON RUBBER BOOM AND DISTORTED ECONOMIC DEVELOPMENT

Bradford L. Barham
Oliver T. Coomes

Dellplain Latin American Studies, No. 34

WestviewPress
A Division of HarperCollins*Publishers*

Dellplain Latin American Studies

Published in 1996 in the United States of America by Westview Press, 5500 Central Avenue, Boulder, Colorado 80301-2877, and in the United Kingdom by Westview Press, 12 Hid's Copse Road, Cumnor Hill, Oxford OX2 9JJ

Library of Congress Cataloging-in-Publication Data
Barham, Brad.
Prosperity's promise : the Amazon rubber boom and distorted economic development / Bradford L. Barham, Oliver T. Coomes.
p. cm. — (Dellplain Latin American studies ; no. 34)
Includes bibliographical references and index.
ISBN 0-8133-8996-8
1. Rubber industry and trade—Amazon River Region—History.
I. Coomes, O. T. (Oliver T.) II. Title. III. Series.
HD9161.A562B37 1996
338.4'76782'09811—dc20 96-8820
CIP

The paper used in this publication meets the requirements of the American National Standard for Permanence of Paper for Printed Library Materials Z39.48-1984.

10 9 8 7 6 5 4 3 2 1

TO OUR PARENTS
FOR LOVE'S OPEN HAND

Contents

Tables and Figures

Tables

Figures

Photographs

Preface and Acknowledgments

After decades of following inward-looking development strategies, many Latin American countries in the 1990s are looking outward to global markets for renewed economic growth and the promise of prosperity. Neoliberal economic policies designed to promote exports of commodities that hold a comparative advantage in global markets—principally natural resources and agricultural produce—are fervently espoused by international financial agencies such as the IMF and World Bank as well as domestic foreign aid groups in the North. Such policies have been adopted increasingly by governments throughout Latin America with the hope of bolstering foreign exchange earnings, improving balance of payments, increasing national incomes, and reducing urban and rural poverty in their countries. Enabled by increasing global integration in markets, finance, transportation and communication, massive changes are underway that flow from this new policy direction: international trade is expanding into non-traditional commodities, industries are being restructured to sharpen international competitiveness, and new frontier lands are being opened up to exploitation.

The recent shift toward more outward-looking, export-based development has prompted much earnest speculation and debate among observers of Latin American affairs as to the potential consequences of neoliberal economic policies. Will such policies increase national income without exacerbating income inequality? How will the poor and disenfranchised groups in society fare? What will be the environmental consequences of such policies, especially in the more biodiverse and ecologically sensitive frontier areas? What should be done to ensure that export-led development may be sustainable over the long term? A rich source of potential insights into these issues lies in the historical literature on past experience in the region with export-led growth. The economic history of many Latin American countries has been marked by the rise and fall of natural resource sectors: for Peru, for example, from the Conquest onwards one can trace out the respective epochs of silver, guano, rubber and more recently of copper, fish, petroleum and coca, all destined for overseas markets. Indeed, a substantial literature has developed around the growing interest in the experience of natural resource extraction in the region; however, much of such work is generally disappointing: accounts often are cast in strongly ideological terms with little appreciation of the underlying economic logic and dynamics of the resource boom and bust phenomenon. Alternate interpretations of the causes and consequences of past resource booms in the region are urgently needed, especially ones that may better inform current discussions of the effects of neoliberal economic policies on the economies, peoples and the environments of Latin America.

The work presented in this monograph on the Amazon Rubber Boom—one of the largest, longest and most volatile of resource booms in Latin American history—represents the fruit of a felicitous collaboration between Barham, an economist, and Coomes, a geographer. A new look was needed, both felt, one that combined the insights of their respective disciplines into a more persuasive, comprehensive, and logically consistent interpretation of the rubber era. This collaboration began in earnest after the latter returned from a year of dissertation field work in northeastern Peru. Having read much of the recent literature on the Rubber Boom and discussed its findings, conceptual approaches and common puzzles, Coomes had departed for Peru where he undertook his research on contemporary livelihood patterns among peasants (*ribereños*) of the Tahuayo River basin, near Iquitos. From interviews aimed at eliciting the environmental history of the river basin he found considerable evidence of local industry-level features that were incongruent with depictions and expectations coming out of recent literature on the Rubber Boom. Upon returning from the field, he began to delve into the classic rubber industry accounts provided by early observers who had lived or traveled extensively in Amazonia during the boom. From such accounts, further evidence was found to suggest that current interpretations were indeed wanting.

As an economist, Professor Barham's interests lay primarily in studying how the organization of natural resource and agricultural activities affected the development process of Latin American economies. He was intrigued by the potential application of recent innovations in the field of industrial organization economics, because they could provide a rich basis for understanding the economic logic of contractual relations, institutional formation and strategic behavior, all of which can play a crucial role in determining how efficiently a sector is organized and how equitably its returns are distributed. In his work on aluminum, bananas and citrus, he had used new theories pertaining to strategic behavior to improve our understanding of certain resource booms in Latin American economic history, but the Amazon Rubber Boom seemed to require working with other recent innovations in contract theory and the microeconomics of risky, decentralized activities. At the more macroeconomic level, Professor Barham's research on current agro-export promotion strategies and their growth and equity impacts had led him to study trade models of booming resource sectors and their potential distortionary effects on the overall economic structure of a boom region.

As a human geographer, Professor Coomes was interested in understanding how economic and social processes mediate the material relationship between rural people of Amazonia and the land upon which they depend for their livelihood. The study of landscape—that dynamic complex of physical and human features on the earth's surface—reveals much to the geographer. Patterns of agricultural fields, forest clearings, or towns and villages, for example, are seen not as "artifacts, accidents or happenstance" but rather as the physical manifestation of underlying economic processes; in a sense, human action renders visible in the land the operation of such processes. Such patterns also are by no means static: the landscape is continually being reworked—often in "fits and starts"—as human needs, technology and trade conditions change. Each "revised" landscape provides an altered set of opportunities and constraints for people who live on the land and thereby conditions resource use decisions and ultimately longer term prospects for economic development. The study of landscape therefore reveals much about essential economic phenomena, especially as one attempts to explicate spatial differentiation and systematic patterning on the land through time.

In the present study, evidence of a highly variegated landscape of property, labor, capital and trade relations across the Amazon basin during the Rubber Boom was of crucial importance. The contrast between upriver and downriver areas was striking: upriver, distant from the major port-cities of Belém, Manaus and Iquitos, were large estates, often worked by desperately poor tappers under a powerful owner who sought to control river trade; near the port-cities were small rubber operations, run by semi-independent tappers who could exchange their rubber for provisions with a diverse group of river traders. Overlooked in previous works, such spatial variegation in the relations of production was inconsistent with the grand theories of the boom and suggested the need for a more sophisticated understanding of the key underlying economic and social processes in operation during the era. Such an understanding was developed using recent economic theories of industrial organization, institutions, and economic change. By asking why certain relations or features prevailed in some places and not others, a stronger test could be made of the robustness of alternate interpretations of economic processes and the most salient legacies of the boom could be identified that conditioned prospects for future development. In this way, the Rubber Boom promised a potentially fascinating case study for both authors that would further our understanding of such processes and the historical landscape of Amazonia.

Our hope in presenting this work as an interdisciplinary endeavor is that others may be persuaded of the worth of building bridges between disciplines, ones that traditionally have tended to keep others at arm's length in the academy. Complex questions, such as those addressed here, call forth the skills and insights of researchers in a variety of fields. By learning the lexicons, ideologies, theories, constructs, and "cultures" of other disciplines, each field is enriched and the ability to solve research problems of more general interest and importance is enhanced. We believe that this work demonstrates how the approach, theory and concerns that exist in the field of human geography can complement those of economics. For the reader who at the outset may be skeptical of the potential benefits of such collaboration, we hope that the rest of the monograph will convince you of its merits and testify to what for us, has been a most rewarding intellectual experience.

We are most grateful to those individuals and institutions that made this study possible. As we implied above, this work emerged in part as a response to the view that the story of wild rubber had been well told and that few worthwhile questions about the era remained to be answered. We are appreciative of those individuals who expressed this view, for it served to motivate us to follow up on our hunches and to take a closer look at the era. But our debt is indeed greater to another group of scholars—those who previously brought the importance of this era to light and provided the foundations upon which an understanding of this period could be built, specifically Barbara Weinstein, (the late) Warren Dean, Roberto Santos, Guido Pennano, José Flores Marín and Stephen Bunker. Although we challenge their views here in many instances, we share their vital curiosity for the boom, and we hope that our work may stimulate the present reader as we were stimulated by their ideas, interpretations, and opinions.

Earlier drafts of this work were presented in a variety of public forums, and we benefitted greatly from the comments made by journal referees, journal editors, and seminar participants. The work presented here in this book is an integration of three journal articles, which appeared in 1994 in the *Hispanic American Historical Review*, the *Journal of Latin American Studies*, and the *Latin American Research Review*.[*] We thank the editors and presses of these journals for their permission to allow us to integrate in one location the arguments developed in the individual articles. This opportunity allows us to achieve two ends that could not be met in journal format. First, we are able to present a more comprehensive and detailed treatment of the Amazon Rubber Boom and its development legacies. Second, we are able also to explore the broader methodological implications of our research for contemporary efforts to examine the logic and role of extraction in economic development in biologically diverse frontier areas. To our satisfaction, this monograph also restores the work with a much improved presentation to its original design as a unified piece that was built up from the minutiae of the organization of the wild rubber industry through the performance of the rubber sector and to broad regional development outcomes.

Of those who directly or indirectly supported this work, we wish to thank in particular the following individuals: Robert Aiken, Anthony Bell, Victor Bulmer-Thomas, Stephen Bunker, Michael Carter, John Coatsworth, William M. Denevan, Cesar Herrera, Sharon Kellum, Linda Kjeldgaard, Eleanor Lahn, Gilbert Merkx, Richard Norgaard, Nelson Pinto, David J. Robinson, Rachel Schurman, and Mark Szuchman. Our appreciation also is extended to seminar participants at the University of Wisconsin (Madison and Milwaukee), University of Chicago, University of Texas at Austin and McGill University as well as attendees at the annual meetings of the Association of American Geographers and Canadian

---

* Barham, Bradford L. and Oliver T. Coomes, 1994. "Wild rubber: industrial organisation and the microeconomics of extraction during the Amazon rubber boom (1860–1920)," *Journal of Latin American Studies*, vol. 26, no. 1, pp. 37–72; Coomes, Oliver T. and Bradford L. Barham, 1994. "The Amazon rubber boom: labor control, resistance, and failed plantation development revisited," *The Hispanic American Historical Review*, vol. 74, no. 2, pp. 231–258; and, Barham, Bradford and Oliver T. Coomes, 1994. "Reinterpreting the Amazon rubber boom: investment, the state, and Dutch disease," *Latin American Research Review*, vol. 29, no. 2, pp. 73–110.

Association of Geographers meetings. The ardent discussion that followed our presentations proved to be most constructive and constituted another of the pleasures of undertaking the work. The views we express here, of course, are not necessarily shared by the aforementioned, nor can they be held accountable for any errors of fact or interpretation that we may offer or any other shortcomings of the work; this liability remains ours alone to bear.

Several institutions provided funding to the authors during the preparation of this work. Professor Coomes received support from the Social Sciences and Humanities Research Council of Canada, World Wildlife Fund, the University of Wisconsin-Madison, and the InterAmerican Foundation for his doctoral studies and dissertation research during which much of the initial work was conducted. A subsequent grant to both authors from the John D. and Catherine T. MacArthur Foundation under the Responses to Global Change Program enabled the preparation, editing, and publication of this monograph. Although the views expressed here are strictly those of the authors, funding provided by these agencies enabled our quest, which may one day benefit both the indigenous peoples and the environment of Amazonia, a result not incongruent with their respective missions. We gratefully acknowledge the support of these agencies.

We thank also our home institutions and colleagues at the University of Wisconsin-Madison and McGill University for providing a welcoming and stimulating environment for research. Our librarians assisted us unflaggingly in seeking out the more obscure materials and by arranging interlibrary loans that brought much needed articles and books to us. Ms. Joanne Lynch provided essential assistance in the collection of research material and secretarial assistance. The fine craftsmanship of the maps and figures was that of Ms. Marcia Harrington at the Department of Geography, Syracuse University. The McGill University PhotoLab produced with great care and attention the photographic plates from the often time-worn and delicate originals. Permission to reproduce Table 4 was graciously granted by T.A. Queiróz (São Paulo).

Finally, we express our gratitude to our respective spouses, Mary and Carmen, for their unwavering support for this extended, collaborative effort and to Mrs. Loretta Hagman, a once-resident and now much missed grandmother, for the many cookies that helped sweeten our marathon work sessions at the dining room table.

Bradford L. Barham
Oliver T. Coomes

Chapter 1

INTRODUCTION

Few areas in Latin America have received as much recent attention over natural resource development or experienced the large number of product booms as Amazonia. With some of the richest and most diverse tropical rain forest in the world, the Amazon basin has been the source of dozens of export products, from *palo do brasil*, waxes, and resins in colonial times, to wild rubber, Brazil nuts, timber, and minerals during this century. Governments have seen the basin not only as a store of untold resource wealth but also as a vent for surplus and a safety value for demographic pressure elsewhere. The consequences of domestic development policies for the rain forest and native peoples of Amazonia are becoming well known and conservationists throughout the world are calling for more sustainable strategies for development of the region. Among the more popular strategies is the promotion of exports of sustainably harvested rain forest products that promise to provide economic benefits to rural peoples and to ensure the conservation of the standing forest. Nevertheless, the potential efficacy and consequences of such an alternate export-based strategy—this one aimed at more sustainable development—are not well known and discussion would benefit from an improved understanding of the nature of past resource booms in the region.

In this monograph we re-examine the Amazon Rubber Boom which reigned for some five decades through the late 19th and early 20th century as the region's first, largest, and longest natural resource boom. Considered to be a novelty item by 18th century travellers in Amazonia, wild rubber would become in the late 1800s one of the most vital and valuable of the new natural resources demanded by the expanding industrial centers of Europe and the United States. With Charles Goodyear's perfection of the vulcanization process in 1839, the number of potential uses of rubber exploded; rubber became an essential input in the manufacture of consumer goods, industrial and commercial machinery, and later military equipment. Between 1860 and 1910, rubber exports would rise to a flood and Pará upriver fine rubber would set the world standard for wild rubber quality. The promise of wealth created by the boom attracted investors, traders, adventurers, travellers, and prospective rubber workers from the world over.

Amazonia became fully integrated into the world economy by international finance, ocean steamers, and telegraph. Champagne flowed as opera troupes played to cosmopolitan audiences in the major urban centers that had been only years before jungle outposts. In 1909, one year prior to the peak of the boom, rubber worth then some $170 million (today equivalent to more than $2 billion) was exported from the Amazon basin. Soon thereafter, another river of rubber—this one rising from British plantations in south Asia—began to flood world markets, sending prices spiralling downward over ten years by more than 95 percent. Amazonian wild rubber could not compete with cheaper and increasingly abundant plantation rubber, and was all but driven from world markets by the 1920s. The longest and most expansive natural resource boom in the Amazonian history was over.

A sizeable and diverse corpus of literature now exists on the rubber era. Historians, sociologists, and anthropologists from both the North and South have been the main contributors, particularly over the past 15 years with renewed interest in Amazonia. In general, scholars have worked independently and from the perspective of one Amazonian country or another to develop particular interpretations of the era that seek to answer a number of the key questions about the boom. Who were the main participants in the wild rubber industry and how did they fare during the course of the boom? What types of rubber were extracted and how? What were the main impediments to modernization of the industry? What were the legacies of the boom? Why did the boom not lead to more balanced and sustained long term development in Amazonia? Little attention, however, has been paid yet by scholars working on the Rubber Boom to the microeconomic logic of the organization of the wild rubber industry, its linkages with other economic sectors, and the investment patterns associated with the boom. As a result, a significant lacuna exists in the literature on the Amazon Rubber Boom. We respond in the present work by applying recent innovations in both microeconomic theory on contractual relations in risky and informationally-constrained environments and in macroeconomic theory on the economy-wide effects of resource booms. Our intention, in doing so, is to both deepen the discussion on the Rubber Boom era and, in the process, develop better means for addressing the more general question of how extractive industries and natural resource booms shape economic development.

In the pages that follow we develop a new and distinctive interpretation of the Amazon Rubber Boom era, one that offers novel insights into the organization of the wild rubber industry at the turn of the century, the legacies of the boom, and why decades of massive economic expansion did not lead to more sustained development in Amazonia. Unlike previous works that focus attention on gross macroeconomic features of the boom, the coercive or contestable social relations of production, or the flamboyant lifestyles and numbing brutality of the rubber barons during the boom, this study seeks to develop a comprehensive portrait of the industry based on a systematic analysis of how the industry was organized at multiple levels during the boom in Amazonia. We offer an account of the specific organizational forms and relations of production seen in the industry, then identify the implications of such organization for surplus accumulation and rationalization of the industry, and finally examine how the macroeconomic environment created by the Rubber Boom conditioned long term prospects for broad economic development.

Our analysis of the rubber industry during the boom builds from the "bottom up," beginning with the basic supply and demand features that shaped the organization of wild rubber extraction and moving upward through the specific relations of production and trade to the macroeconomic environment created by the boom in rubber. Along the way, we explicate two of the most controversial features of the industry—debt relations and price gouging in trade—by referring to microeconomic factors such as risk, factor scarcity and mobility, and transaction costs entailed by rubber extraction in this environment. Such factors also are shown to have influenced strongly the potential for local surplus accumulation and transformation of the industry to plantation-based production, both central issues in the literature on the boom. With empirical evidence of significant surplus retention by local participants, we shift our attention away from the performance of the rubber industry and toward intersectoral patterns of investment to show how the boom in rubber created a macroeconomic environment that brought dramatic economic expansion but also limited long term prospects for sustained development by the distortions it engendered in the economic structure.

The monograph is divided into two parts, each containing five chapters. In Part I, we examine the organization and performance of the Amazonian wild rubber industry during the boom. In Part II, we assess the macroeconomic patterns of private and public investment generated by the Rubber Boom, and explore the economic and political logic underlying why these patterns evolved as they did toward a highly fragile economic structure. The two parts of the monograph are linked principally by the finding in Part I that substantial local surplus retention and accumulation in the wild rubber sector gave rise to the patterns of investment and development in the regional economy which left the economy highly vulnerable to a collapse in rubber prices.

Part I begins with Chapter Two and a review of how prior analysts have understood the organization of the Amazonian wild rubber industry. Three purported problems with the performance of the wild rubber sector are highlighted; previous researchers generally conclude that social relations or ecological conditions held back the sector's performance. Those problems were:

- *Why was the supply of wild rubber so unresponsive to variations in world market demand?* Despite burgeoning demand with rapid technological innovation in Europe and North America, the supply of rubber from the Amazon fluctuated wildly—from a flood to a trickle and back—bringing price instability that frustrated both overseas buyers and Amazonian producers.

- *Why were labor and trade relations so resistant to reform?* Analysts since the boom have pointed to credit-debit relations and the highly decentralized and monopolistic trading system as one of the major impediments to rationalization of the industry.
- *Why were rubber plantations not successfully developed in Amazonia, particularly in the face of ferocious competition from Asia?* Plantations seemingly promised high returns from intensive and orderly production of rubber, in contrast to the apparently inefficient, artesanal manner by which wild rubber was collected from the Amazon rain forest.

Our attention thus focuses first on how previous analysts understood the structure of the industry and the key impediments to further expansion of production. A critique is offered of prior analyses and interpretations—both from the turn of the century and more recent contributions—one that points to the need for a fresh look at the industry's organization and performance during the boom. To enable a more fruitful re-examination, we also suggest an alternate analytical approach for studying the wild rubber industry that extends classic industrial organization theory beyond the usual confines of "structure, conduct and performance" to include recent conceptual innovations from neo-institutional economics, particularly in the area of the role of risk and transaction costs in shaping contractual relations. This extension promises to be of potentially significant value to contemporary efforts to understand the role of extractive activities in forest peasant household economies and how they can be pursued in ways that enhance the prospects for more sustained development.

The core of our organizational analysis of the Amazonian wild rubber industry is presented in Chapters Three to Five. In Chapter Three, we examine the basic characteristics of the rubber industry, specifically the nature of markets, extractive activities and trade. Our analysis opens with an overview of the international rubber trade during the late 19th and early 20th century. The prime position of Amazonian wild rubber on world markets and the fundamentally competitive—not monopolistic—nature of the rubber trade suggest the importance of surplus generated by rubber extraction and available for local investment over purported surplus drainage or foreign monopoly control of the industry. The magnitude of such returns and how they were distributed across industry participants are shown in subsequent chapters to be strongly influenced by the organization and structure of the industry. Chapter Three also introduces the important risk and transaction cost features associated with the decentralized and dispersed extraction of wild rubber in a jungle environment.

In Chapter Four, we turn to the conduct of the wild rubber industry. Our analysis here focuses on the relative scarcity of key factor endowments—abundant but dispersed wild rubber trees, scarce labor and initially scarce capital—as well as the risks, mobility of labor and capital, and transaction costs inherent to working in the Amazon rain forest during the late 19th century. How these features in turn shaped property relations, labor arrangements, relations of capital and exchange is central to understanding the organization of the industry and the dominant contractual forms that linked the various participants, from rubber gatherers to estate owners, river traders, and urban suppliers. These relations of extraction and trade, in turn, influenced the level and distribution of returns received by the various actors in the industry. Geographic variations in contractual structures and the distribution of returns are also explored, as distinctive features in different parts of the vast Amazon basin gave rise to variations in risks, mobility, and transaction costs and hence industrial organization and performance.

Our revised understanding of the wild rubber industry allows us in Chapter Five to reassess the three aforementioned problems in the rubber sector's performance believed by previous scholars to hold the answer to the greater problem of why the boom did not lead to more sustained development in Amazonia. Prevailing explanations for each issue are assessed in turn and, in each case, an alternate explanation is offered based on our industrial organization analysis. Moreover, the relevance of such sectoral performance issues in understanding the broader issue of unsustained economic development in the region is seriously questioned. The bottom-up approach to understanding the organization of the wild rubber sector is succinctly summarized in Chapter Six with three goals in mind: (1) to draw out the key findings from in the previous chapters; (2) to explore the conceptual contributions achieved in that analysis; and, relatedly, (3) to highlight the general lessons for future analyses of extractive industries. The conceptual contributions and general lessons are called on again in the closing chapter of the book.

Part I concludes with two controversial findings. First, the wild rubber industry during the boom was not relatively inefficient as observers both past and present assert and, hence, significant transformations in the social relations of the activity—whether it be the introduction of wage labor or other tapping contracts—would not have yielded lower-cost wild rubber, given the local conditions and the distinctive endowment and informational advantages of tappers, patrons, and traders. Second, the industry was not plagued by a lack of locally retained surplus, purported to have foreclosed opportunities for economic diversification and longer-term development following the boom. Local surplus retention was, in fact, abundant, especially in the major port-cities of Belém, Manaus, and Iquitos, and available in varying measure to participants at all levels of the industry for investment in the developing boom economy.

The focus of our analysis shifts in Part II from the rubber industry to the entire macroeconomy of Amazonia during the boom in an effort to better understand the crucial question of why decades of growth in the rubber sector did not lead to broader, more sustained economic development in the region. We begin the analysis in Chapter Seven by critically reviewing current macrolevel explanations for the failure of the Rubber Boom to promote long term development in Amazonia, including dependency theory, the political ecology of underdevelopment, and Marxian theory of thwarted development. Each provides a distinct perspective on the problem of unsustained development and all three rely upon the assumption of limited surplus accumulation and/or the assertion that the industry was fundamentally inefficient. Our findings from Part I challenge these assumptions and suggest a novel direction: to follow both public and private investment of rubber surplus to see how the macroeconomy was affected by the booming rubber industry. To do so, we adopt a theoretical model developed by the Australian economists W. M. Corden and J. P. Neary, and applied by others to understand the perverse effects of a natural resource boom on the structure and performance of a small, open economy. Often referred to as the "Dutch Disease" model, after the experience of the Dutch with their natural gas economic boom during the 1970s, the model is well suited to explain patterns of investment observed during the Rubber Boom and to explore their consequences.

In Chapters Eight and Nine, we employ the Dutch Disease model to focus our analysis of private and public investment, respectively, in Amazonia during the boom. The nature of private investment within the rubber industry is considered from investment in establishing rubber estates and providing tappers with the necessary provisions to the establishment and operation of the trading system that linked the hinterland estates with rubber markets overseas. We find that much of private investment was concentrated as working capital which was highly specific to rubber extraction, and which generated few downstream or lateral linkages. Moreover, opportunities for investment in the production of other tradeable goods (e.g., agricultural products) were diminished by the boom as investment and re-investment was concentrated increasingly in the booming rubber industry and the non-tradeables sector (e.g., real estate, trade facilities, construction, tertiary activities, and consumption). In Chapter Nine we then follow investment by Amazonian governments of the abundant surplus captured from the rubber trade through import and export duties, and find that patterns of public investment likewise were affected by the boom, with state expenditures also following rubber and the non-tradeables sector. The bulk of state revenues captured from the rubber industry was not directed towards industrialization, diversification or economic reform but rather to securing territorial claims and concessions, regional articulation and administration, and state subsidies and grants. Powerful incentives created by the boom shaped the political economy of Amazonia increasingly toward rubber. At the peak of boom, such investment had created an unusually fragile macroeconomic structure, one that balanced precariously on the world price of wild rubber. When prices collapsed, no other resource or activity could provide the same level of economic returns as wild rubber and the region spiralled into recession, leaving behind the vestiges of a once more prosperous time.

We conclude in Chapter Ten that in the case of wild rubber, such a macrolevel analysis, when founded upon a clear and explicit understanding of the microeconomic logic of industrial organization, provides a more persuasive explanation of the boom and its aftermath than other previous explanations. Here, we summarize the most salient findings of our research on the Amazon Rubber Boom. New directions for future research are identified toward the resolution of the outstanding empirical issues raised by our analysis of the rubber industry and the boom era.

In Chapter Eleven, we set forth the underpinnings of an analytical framework for studying the nature, contributions, and problems of natural resource extraction, especially forest products, in economic development. One of the fundamental questions that structures the approach within the framework is whether the extractive activity is in a boom period like the historical case of Amazonian wild rubber, where returns to this activity dominate alternative tradeable goods, or whether the extractive activity is more of a complementary activity, where returns are not great enough to dramatically alter production activity in other tradeable products. In the case of the former, emphasis would lie on examining the industrial organization of the boom sector to understand how the geographic and market features as well as the choices made by economic agents combine to shape efficiency and equity outcomes. In the case of the latter, our attention turns to exploring the household's choices across distinct economic activities and the conditions under which they would tend to rely more or less on forest extraction as part of their economic livelihood. In both cases, a host of important economic and geographic questions are then raised which, upon further study, are likely to show how contemporary extractive activities in the developing world contribute to the process of economic development and human-induced environmental change.

PART I

The Organization And Performance of the Wild Rubber Industry In Amazonia

Chapter 2

MODELS OF THE INDUSTRIAL ORGANIZATION OF EXTRACTIVE ACTIVITIES

Introduction

The fate of the rain forests and traditional peoples of Amazonia has emerged over the past two decades as one of the most pressing issues on the environmental agenda. Popular accounts, academic treaties and conservation literature on the development challenges in the region all point to the need for an improved understanding of the multiple ways in which the rain forest's resources are used by local peoples. Traditional resource use practices and activities are seen as possibly offering an alternate basis for more effective and sustainable forest management schemes, ones that both provide for the conservation of rain forest resources and serve to improve the welfare of rain forest peoples. Such interest has stimulated considerable recent research on the nature of traditional extractive practices, such as the gathering of non-timber forest products, fishing, and hunting. In addition, scholars are looking back in time at past extractive activities and eras, such as the Amazon Rubber Boom, with an eye to better understanding the historical evolution of economic cycles of extractive activity and the role of extraction in regional development.

One of the most striking features of recent research on current resource extraction in Amazonia is a pervasive tendency to consider traditional use essentially as non-economic (or minimally economic) behavior. Most work to date on traditional resource use is descriptive; nature-rich narratives are offered of the techniques, technologies, and management practices employed by specific groups as well as some indication of the attendant ecological benefits. Although some studies do also provide a sense of the financial returns from such use (e.g., cash income), few consider the more fundamental problem of how forest peoples make strategic choices about resource use and the role of extractive resource use in their household economy. This shortcoming is critical, because efforts to develop policies and institutions to promote more sustainable resource use are not generally informed by a strong understanding of the microeconomic logic of local forest resource use.

In several ways, the Amazon Rubber Boom provides a rich takeoff point for scholars and practioners interested in developing a more systematic framework for addressing sustainable development concerns. For one, rubber extraction was studied rather exhaustively during the boom era, mostly by industry analysts who were trying to determine how to improve the productivity and regularity of supply of wild rubber to the rapidly expanding manufacturing centers of Britain, Europe, and the United States. Their writings provide much information on the organization and performance of the rubber sector. A second reason for starting with the Rubber Boom lies in the allure of rubber extraction as a sustainable resource use activity. Wild rubber can be extracted from the most common species of rubber trees in a sustainable manner, and as such, the Amazonian rubber industry (both past and present) has become the focus of much attention among the conservation and development community. Of additional interest to development analysts is the dominant position held by the wild rubber industry in Amazonia during the boom. The task of understanding the development process of a region dominated by an extractive activity is somewhat simplified by the dominance of one, rather than multiple sectors. Finally, the failure of the Rubber Boom to sustain long-term economic development in the region raises a number of questions about what went wrong in the sector, and about the potential of resource extractive industries to propel regions onto more sustained growth and development paths. For all of these reasons, the Rubber Boom provides a propitious point of departure for the exploration of models of extractive activities in the Amazon.

We begin this chapter with a review of the literature on the Amazon Rubber Boom. Our aim here is to show how prior writers understood the organizational logic and economic dynamics of the wild rubber industry. We divide our review in two parts, first considering the works of early observers—those written during the boom era—and then more recent studies undertaken by scholars of the boom. We develop our review with a particular eye to the ways in which previous approaches may or may not contribute to the development of a more general framework for understanding the organization of resource extractive industries. Our review reveals a diversity of perspectives on the organization of the industry and suggests the limited potential of previous approaches, taken separately or together, to provide the foundations for a viable analytical framework. We turn therefore to modern industrial organization theory and examine how such theory may be useful for the study of extractive industries in a developing economy. We conclude the chapter by presenting the outlines of an analytical structure that serves to guide our path of inquiry on the wild rubber industry in the remainder of Part I.

Previous Approaches to the Amazon Rubber Boom

Early writers on the Amazon Rubber Boom were concerned primarily with two issues, both related to the performance of the rubber sector. The most pressing issue during the boom era itself was the pragmatic problem of how to improve the supply of wild rubber. Most observers argued that the rubber supply problems were due to the inefficiency of traditional methods of rubber collection and pointed to the need for industrial reorganization or wholesale transformation of the wild rubber industry to plantation-based production. The second issue, which arose very late in the boom, related to the social condition of workers engaged in the extraction of rubber, especially in some of the more remote areas dominated by "rubber barons". Considerable concern was raised then over the social impacts of wild rubber collection on Amazonian peoples. More recent scholarly work shares this earlier social concern, but also seeks more broadly to understand why sustained economic development did not result from the boom. By reviewing how prior writers addressed and judged the performance of the wild rubber industry, we can both explore the ways in which the organization of the sector has been explained previously and identify some key features of the sector that would need to be integrated into our analytical framework.

Early Writings

At the close of the 19th century, world demand for rubber—a veritable "miracle product" of the time—pushed prices upward, and the Amazon basin was recognized as the world's primary source of the finest grades of wild rubber. The region was visited by waves of industry men, trade consular officers, government officials, explorers and casual travellers seeking to know more of the nature of the booming rubber trade in Amazonia.[1] The finely descriptive literature from the era not only provides a window on the organization and operation of the rubber industry during the boom, but also signals the apparent impediments, as well as possible remedies, to further expansion and development of the industry. In retrospect, it is understandable that these works drew heavily upon the experience, intuition, and common sense of their authors for insights into the rubber trade rather than a systematic framework for analyzing the organization of an extractive industry. Microeconomic theory and, particularly, methods for the study of industrial organization were only in early stages of conceptual development

---

[1] Their observations are recorded at length in a variety of reports (e.g., Akers, 1912; Labroy and Cayla, 1913; Pearson, 1911; Schidrowitz, 1911, trade journals (e.g., *India Rubber World, India Rubber Journal, South American Journal*), consular and trade reports (e.g., Plane, 1903; *British Consular and Trade Reports*; U.S. *Daily Consular and Trade Reports*), regional domestic studies (e.g., Ballivián and Pinilla, 1912; Nery, 1901; von Hassel, 1905; Fuentes, 1908; Walle, 1907, 1914; Maúrtua, 1911; Salamanca, 1916; LeCointe, 1922), and travel accounts (e.g., Ordinaire, 1892; Enock, 1910; Lange, 1912,1914; Woodroffe, 1914).

at the turn of the century; the first major wave of industry studies came only in the 1920s and 1930s.

The most common and persistent concern of early writers was the high cost and related unresponsive (inelastic) supply of wild rubber.[2] In Europe and North America, rapid innovation led to the incorporation of rubber into a wide variety of industrial equipment and consumer goods. As demand surged in the latter decade of the 19th century and first decade of the 20th century, the harvest of wild rubber increased only haltingly, and the world experienced periodic shortages of crude rubber (Coates, 1987:137). Rubber became an notably expensive commodity, both in absolute terms and relative to other raw or processed materials used in manufacturing during the boom era. In 1908, for example, prior to the peak in rubber prices, one ton of rubber cost manufacturers about $3,000 whereas steel cost $3.30/ton (*India Rubber World*, vol. 38, no. 1, April 1, 1908, p. 209).

The combination of surging but cyclical demand, inelastic supply, and fears of someone "cornering" crude rubber output—a product for which no substitute existed—caused rubber prices to fluctuate widely. Data from LeCointe (1922:I:431) suggest just how volatile rubber prices were; during the period between 1890 and 1910, the annual high price exceeded the low price for fine rubber at Belém by more than 25 percent in all but three years and by more than 50 percent in 11 of the 30 years. Such volatility prompted Colonel Samuel Colt, then president of the United States Rubber Company, to remark about crude rubber, "[w]e are not dealing with a commodity. We are dealing with dynamite." (Lawrence, 1931:2). Price instability carried major risks for foreign manufacturers who needed a steady and secure supply of rubber, as well as for Amazon rubber entrepreneurs, who outfitted workers to collect rubber that often only would be delivered several months later, at the end of the season. The challenge, to exporters and major industrial consumers, was thus to increase supply responsiveness so that crude rubber stocks would better match industrial demand, thereby relieving the growing shortage and bringing stability to the consumer market and the Amazonian economy.

The main solution to the crude rubber supply problem in the minds of industry experts at the time lay in the rationalization of rubber production. The traditional manner of procuring rubber in the Amazon—collection from the wild by widely dispersed tappers—was seen as backward and inefficient. Similarly, the costs and risks associated with provisioning the tappers and moving the wild rubber over large distances, frequently on perilous rivers, was also viewed as a major obstacle that could be overcome by modifying the practice of rubber gathering and trading. Industry analysts argued for a variety of improvements, including the promotion of new methods and

---

[2] The problem of supply inelasticity would resurface during World War II when the U.S. Rubber Development Corporation sought with limited success to increase the output of Amazonian wild rubber as a means to offset the loss of access to southeast Asian rubber plantations to the Japanese (see Higbee, 1951).

technology for harvesting rubber, elimination of the numerous intermediaries that stood between the extractor and the market (e.g., exploitative patrons and monopolistic traders),[3] upgrading transportation facilities and infrastructure (Hale, 1913), and a reduction in the cost of provisionment (Russan, 1902) as well as the threat of disease to the rubber worker (Cruz, 1972).

The most profound reform suggested for the rubber industry was wholesale transformation, from one based on decentralized wild extraction to an industry centered on plantation-based production. The American view on the need for plantations was perhaps best expressed by Colonel Colt in his remarks to the annual meeting of the New England Rubber Club on February 24, 1905,

> It is to the Para of the Amazon valley that we must look for the permanent solution of the crude rubber problem. We merely must devise improved methods to obtain the rubber. By systematic development and effort the production of Para rubber can be established on a permanent basis which will give it a position among raw materials practically as reliable as cotton or corn. (*India Rubber World*, vol. 31, no. 6, March 1, 1905, p. 204).

The British, of course, had quietly turned to the development of rubber plantations in south Asia in the 1870s (though significant planting began only in 1898 in Malaya and 1899 in Ceylon), and in 1905 the first shipments of commercial quantities—some 176 tons—of plantation rubber arrived in London, realizing exceptional profits (Lawrence, 1931:13–14).

The problem of shifting the wild rubber industry to plantation-based production took on new urgency for Amazonia in the first decades of the 1900s, with ever-growing quantities of low-cost rubber from Asian plantations flooding the world market. A fundamental shortage of low-cost labor in the Amazon basin was seen as the primary obstacle to transforming the industry, and some analysts recommended the recruitment of Asian laborers, coupled with state fiscal incentives and technical assistance, to reduce the labor constraint and promote plantation development in the Amazon (see Akers, 1912; Oyague y Calderón, 1913; Woodroffe and Smith, 1915). Although regional and national governments throughout the basin would never sponsor massive Asian immigration, they did provide a variety of development incentives, the most notable of which are found in Brazil's "Rubber Defense Plan" of 1912 (see Weinstein, 1983b:225–229). Nevertheless, the Amazon rubber industry would not be transformed despite ruinous competition from Asian competition. Even today, after frustrated

---

[3] Most experts focused on improving tapping tools and techniques, though at least one writer argued for the elimination of collectors themselves, to be replaced by electrical wires that would stimulate the flow of latex on demand (see von Hassel, 1912). Woodroffe and Smith (1915:137–138) considered the Amazon rubber trade ("truck") system itself to be "nothing more than organized robbery on a huge and particularly cruel scale" (p.137) and argued that the industry would never be sound commercially until the truck system was abolished.

attempts by Henry Ford in the 1930s and multinational tire companies in the 1960s, the Amazon rubber industry continues to be based primarily upon the collection of wild rubber from the rain forest.[4]

As the Rubber Boom faltered in the 1910s under competition with Asian plantations, a second corpus of literature began to emerge that would profoundly shape popular understanding of the era. Hardenburg's (1912) account of the extreme cruelties and exploitation of native peoples on the vast Putumayo estates of J. C. Arana and the Peruvian Amazon Company shocked the world and, with confirmation by the British Casement Commission, gave a new and sinister face to the wild rubber trade in Amazonia. The Putumayo scandal initiated a longstanding concern over the impact of the trade on the social condition and culture of Amazonian native peoples (Murphy and Steward, 1956; San Román, 1975; Ross, 1978; Chirif and Mora, 1980:279–301; D'Ans, 1982:159–196; Hemming, 1987:271–314)[5] as well as a fascination with the rubber barons and indebted tappers (Wolf and Wolf, 1936; Ferreira, 1934; Reyna, 1942; Collier, 1968; Fifer, 1970; Muratorio, 1991). Many of the more popular works have left a haunting impression of the Amazon rubber trade as one dominated by large estate holders who coerced, enslaved or debt-bonded workers to gather rubber to be sold to serve the barons' excessive appetites for conspicuous consumption. Richard Collier (1968:55) writes of the barons,

> [s]uch men were monarchs more absolute than Eastern kings; to enter their water-highways was akin to piracy, and the limits of the trails lying behind those highways had been charted with blood.

Where enslavement or overt force were not used, debt peonage was argued to hold workers in the forest, where they tapped the rubber trees and slid ever deeper into misery. Howard Wolf and Ralph Wolf (1936:34) describe "... the vicious credit system which later [during the boom] made an actual slave of nearly every rubber gatherer, whether technically slave or free." This view of social relations during the boom persists today. In Coates' (1987:95–96) view,

> [d]ebt became the opiate of the seringueiro. Belém's trade arrangements moved from the picturesque to the macabre—the mentality of the small shop run wild. As rubber sky-rocketed, men from other parts of Brazil streamed into the provinces of Amazonas and Pará, to become debt-slaves in their turn.

Another authority writes, "[i]t was this control over the means of production, that is, the land and the trees that grew on it, and the labour process that enabled the seringalista to force such oppressive rates of exchange on the

---

4 In 1985, Brazilian industry consumed some 405,000 tons of rubber products of which only 2% was derived from planted Brazilian rubber trees (Dean, 1987:163).

5 Concern over the treatment of Amazonian native and mestizo people involved in extractive activities remains today. See for example Burkhalter and Murphy (1989), AIDESEP (1991) and Romanoff (1992).

seringueiro and keep him captive on the seringal."(Bakx, 1988:148). Only recently have a small number of scholars begun to reconsider such views (see Weinstein, 1986; Taussig, 1984; Gray, nd.).

From the early accounts of the Rubber Boom, the basic technology and form of contractual relations which defined the industry are evident. Simply put, widely dispersed individual tappers using simple equipment to extract wild rubber, frequently under contractual ties to traders and patrons who provided provision advancements to enable the extractive activity to occur. We note several other distinctive features of the industry in these accounts, including:

- the relatively high cost of labor in much of the Amazon (wage levels in the Amazon were several times higher, for example, than in Asian countries, where plantation rubber cultivation was developed);
- the mobility of labor (tappers frequently migrated from the rubber estates to major river port towns during the rainy season);
- the prevalence of itinerant river traders along many rivers, trading rubber for cash and goods;
- the multiple risks (i.e., in nature as well the market) associated with tapping and trading activity in a highly dispersed environment;
- the prevalence around the major port-cities and the mouth of the Amazon of semi- or fully independent tappers (often called *caboclos*); and,
- the labor supervision difficulties experienced by foreign investors attempting to establish more intensive wild rubber extraction operations, especially under wage payment systems.

The richness of these observations is not mirrored, however, by early analyses that attempt to account for why the decentralized form of industrial organization predominated over other possible alternatives. Indeed, only Akers (1912) and Schurz *et al.* (1925) of the early writings provide a relatively clear portrayal of how the underlying factor endowments in the region (i.e., scarce labor and abundant trees) might have limited the options for plantation cultivation, but little to no attention is given to other explanatory factors, such as risk, the potential costs of monitoring of labor, or other features of the industry that might have accounted for why a rationalization into plantation production did not occur despite repeated efforts by industry participants and regional governments.

RECENT WORKS

The 1970s and 1980s witnessed a revival of popular interest in Amazonia, as concern over the region's forests and peoples grew. As noted above, one important element in this revival of interest has been the development of a body of scholarly literature on the Rubber Boom era that seeks to understand why such expansive economic growth did not beget sustained development. In answering the question, recent works generally adopt (explicitly or implicitly) the primary criticism of early trade observers—that wild rubber extraction was an inefficient form of production—and see the region-wide failure to transform the industry as the principal impediment to development in the era. According to this view, technological change and hence economic development were thwarted by a lack of sufficient surplus for investment in plantation production either because traditional (inefficient) extraction generated minimal local surplus and/or most surplus was drained out of the region to Britain, Europe, the U.S. or the capital cities of Brazil, Peru, and Bolivia by way of unequal relations in local and international trade.

In considering the impediments to rationalization of rubber production, recent studies emphasize the nature of social relations that defined the traditional system of rubber collection, specifically the apparent exploitation of tappers by patrons and traders through the debt contracts. Patrons and traders are argued to have held a position of privilege in the industry, monopolizing trade and holding workers in bonds of perpetual debt, first marking up the tappers' provisions and later discounting the price paid to tappers for their rubber, so that the tappers received little more than subsistence returns. Patrons and traders, motivated by the need to monopolize trade, are viewed as instrumental in resisting or thwarting the introduction of new forms of social relations, as well as new technologies, that might have transformed the industry and permitted balanced economic development in the region.

Two explanations have emerged in the literature for why the traditional social relations persisted during the boom; both are based on the relative power of patrons and traders versus rubber tappers. The classic view of tappers as debt peons, serfs, or slaves suggests that relations were maintained by force or coercion through threat, debt, or violence. Patrons and traders whose profit depended on exploiting rubber tappers saw little to be gained by passing on to them the benefit of price increases, or by altering relations of extraction or trade according to the wishes of a foreign company purchaser. This coercive view of capital-labor relations is reinforced by accounts like that of the British Casement Commission, where, in certain instances and places, rubber barons appear to have exerted considerable power over indigenous groups.

Barbara Weinstein (1983a, b; 1986) argues, by contrast, that tapper resistance—not coercion—conditioned the nature and durability of social relations. Rubber workers, in Weinstein's view, preferred the autonomy of wild rubber extraction over the drudgery of wage labor, and managed effectively to resist proletarianization. Their resistance was abetted by two important features in the industry. First, the extracted rubber was, by law and all practicality, the property of the tapper and not the estate owner who held the land and the trees. Second, the high cost of monitoring the working tapper strictly limited the degree to which patrons could effectively control their workers or curb illicit rubber sales. A durable alliance was thereby formed between tappers and traders, based on interlocking self-interest—the trader's need to control exchange and the tapper's preference for autonomy—that effectively frustrated local and foreign pressures for rationalizing rubber extraction and trade.

Potentially the most significant way to rationalize the Amazon rubber industry would have been the introduction of large-scale rubber plantations similar to the Asian model. Today's scholars offer three distinct explanations for why this effort failed. The first, rooted in dependency theory, and mentioned above suggests that foreign imperial powers (as well as national governments) drained away the local surplus through unequal exchange and, with the rubber seeds they also spirited away, developed plantations in Asia that undercut Amazonia's position in world markets (e.g., Bonilla, 1977; Santos, 1980; Haring, 1986; Flores Marín, 1987). Foreign firms and governments, according to this view, deprived the region of capital for investment in plantations and were therefore the principal culprit.

Warren Dean offers the second explanation, partly in response to the dependency view. South American Leaf Blight, which flourished in the closed-canopy Amazonian rubber stands, reduced yields so drastically that plantation production became uneconomical. Of initial plantation efforts, Dean writes,

> The real problem ... was the decline in labor productivity that was directly the result of the blight. As many of the trees were killed off, and as the survivors provided poor yielders, laborers lost interest in a form of field work that was relatively unremunerative (Dean, 1987:66).

Brazil's 150 year "struggle for rubber" has been essentially a long-running contest with an ecological problem that even today remains a serious impediment to the viability of plantation rubber in Amazonia.

Weinstein's view of labor relations (Weinstein, 1983a,b; 1986) can be extended to a third, Marxist explanation: plantation development was effectively blocked by the durable tapper-trader alliance. Numerous trade journal articles record the dismal failures of foreign firms operating wild rubber estates and point to disappointing yields—well below historical norms—as the cause. Ashmore Russan (1902:7) provides us with a striking example in the experience of the Brazilian Rubber Trust, which could neither realize historical levels of production nor eliminate the credit system. Whereas under Brazilian management the properties had yielded about 250 tons per year, under English direction average production never exceeded 60 tons per year and averaged some 50 tons per year. In this case, the difference was attributed to tappers' sales to "rubber pirates"; in others, lower yields were thought to be due to tappers working at a slower pace. The bond that joined tappers and traders was formed on the basis of mutual self-interest: tappers controlled production but not exchange, and traders controlled exchange but not production. Any effort to push tappers to work more intensively would diminish their autonomy and provoke their resistance; in a sense, the low density of wild rubber trees guaranteed tappers a degree of autonomy they could not realize on a plantation.

Traders who captured surplus through monopolized exchange, presumably also would have resisted the rationalization of production, but for a different reason: plantation development would reduce the need for long-distance river trading (because land close to the cities would be used more intensively, and estate owners might then take up product transportation). This change would diminish traders' opportunities to monopolize trade and thereby lower their returns. In Marxist terms, the precapitalist formations on which the rubber trade was based proved highly resistant to change; they hindered the adoption of more efficient labor arrangements (that is, wage-based production), limited the potential for local capital accumulation, and therefore prevented transformation of the extractive economy.

As this review demonstrates, classic and recent writers alike have sought to develop an understanding of the era and to answer several specific, development related questions by examining the nature and organization of the rubber industry. Although we agree with the importance of focusing on the rubber sector, we would suggest that neither group has provided a sufficiently sharp portrait of the Amazon rubber industry to adequately explain why supply was so inelastic, why social relations and basic industrial organization proved so durable, or why plantation development was thwarted. Classic studies predated developments in formal methods and theories of microeconomics and industrial organization, and more recent works tend to emphasize the nature of social relations over a systematic study of the microeconomic logic of rubber extraction and trade. None of the works integrate an understanding of the basic nature and influence of factor supply, markets, technology, or risk; the intersection of markets and relations of production; or how such relations are affected by basic market processes.

An Industrial Organization Approach

In this section, we introduce the reader to industrial organization theory and move toward a framework for analyzing the Amazonian wild rubber industry. We begin by describing what has become known as the classic industrial organization framework, referred to generally as "The Structure, Conduct, and Performance Model" (see Scherer, 1990; Shepherd, 1985). After reviewing the basic features and logic of the SCP Model, we turn to consider a particular set of recent advances made by researchers in applied microeconomics, modern industrial organization theory, and development economics. We emphasize the key insights contributed by scholars such as Stiglitz (1986), Carleton and Perloff (1989) and Bardhan (1989) which serve to improve substantially our capabilities to analyze and understand the nature of economic activities where information is costly to obtain and risk is prevalent. This broad conceptual discussion is then directed at the problem

of understanding the organization of resource extractive i.
developing economies, and used to tailor our framework for the
the Amazonian wild rubber industry.

THE CLASSIC STRUCTURE-CONDUCT-PERFORMANCE MODEL

The classic industrial organization model was developed to account for the performance of individual industries or sectors. Analyses of industrial organization based on the model typically begin with the basic characteristics of the industry under study, and then move on to account for the market structure of the industry, the conduct or behavior of firms in the industry and then to industry performance (see Figure 1). The model's main explanandum has been to identify the extent of competitive performance in particular industries, and as such has made it an important tool in anti-trust efforts. Industry performance is measured according to a suite of indicators, including the competitiveness of product pricing, the efficiency of production methods, the level of profitability, and the prevalence of technological innovations in production and quality of the product.

In conceptual terms, the structure of an industry is seen to be shaped by certain underlying supply and demand characteristics of the industry (together with government regulations). Among the most influential industry characteristics are the demand-side attributes of the product, the availability of key raw materials and factors of production, and the production technologies. If an industry, for instance, is based primarily on a scarce raw material, and high capital costs are entailed in entering into production at an efficient level, then few firms are likely to dominate the industry. The aluminum industry, with its reliance on bauxite and large amounts of cheap electricity and the large-scale economies in alumina processing, would be one such industry.[6] Industry structure is characterized not only by the number of firms but also by the barriers to entry that limit competition, the size of the markets, and the rules governing trade between distinct regional or national markets. Moreover, industry structure is seen strongly to determine the conduct of firms, which can range from atomistically competitive to singularly or cooperatively monopolistic behavior. Where firms are few and dominate the industry, for example, the structure of the industry is more concentrated, and firms are more likely to behave like monopolists, setting high prices (i.e., above marginal costs) and reaping large profits. Such performance outcomes at the industry level, however profitable to individual firms, are considered to be inefficient.

The classic SCP model is rather useful in the case of wild rubber. In reflecting upon the early and recent writings on the Amazonian wild rubber industry, we note a particular lack of attention to the basic supply and

---

6 The particular historical case of the international aluminum industry is explored by Peck (1988) and Barham *et al.* (1994).

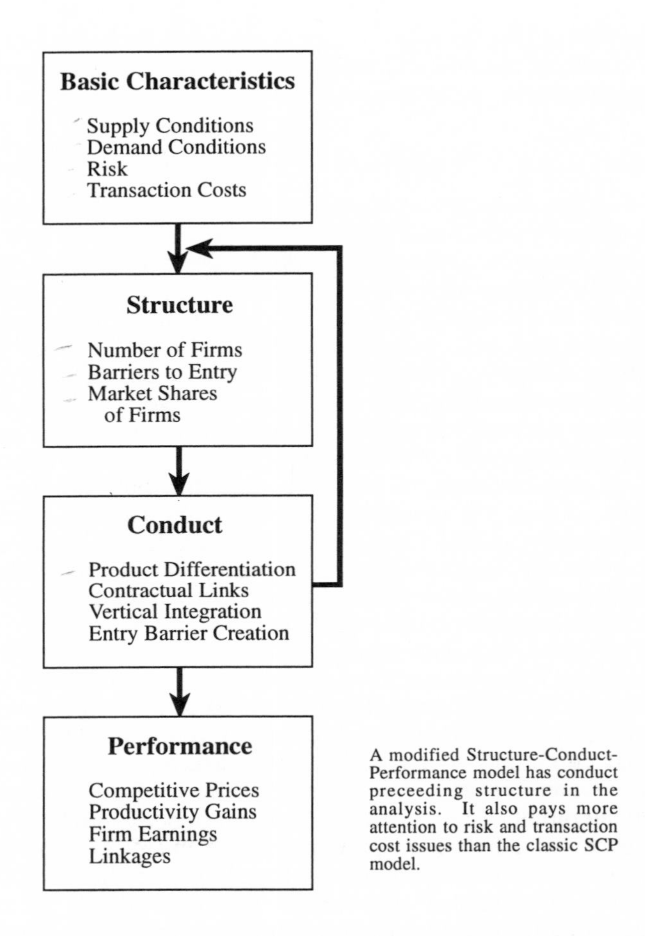

Figure 1. The Structure-Conduct-Performance Model of Industrial Organization

demand characteristics of the industry, and how such characteristics systematically influenced both the structure of the industry and the conduct of its participants. In most works, the conduct of rubber industry participants is largely taken as given or connected only loosely to perhaps one or two basic industry characteristics, prior to discussing performance outcomes. In other words, the arguments found in the Rubber Boom literature regarding performance problems, such as monopolistic trading, inefficient technological choices, or plantation investment failures, are generally derived from an incomplete picture of the underlying supply and demand characteristics of the industry. Our aim in Part I of this monograph is to build up a more theoretically informed and disciplined analysis of industry performance on the basis of a clear appreciation of the underlying supply and demand characteristics, the structure of the industry, and the conduct of rubber industry participants.

Recent Innovations in Industrial Organization Models

Over the past three decades, the classic SCP model has been revised in two major ways that make the model more amenable for use in the study of resource extractive industries. First, the notion that firm conduct is determined exclusively by industry structure has been challenged, and replaced with the view that conduct of participant firms may be determinative of structure (as well as vice versa) (see Figure 1). In the case of the aluminum industry, if prime bauxite reserves and cheap hydroelectric sites are, in fact, scarce, then early entrants to the sector could (and Barham, 1994 argues they did) act strategically to secure excess capacity in these holdings to prevent further entry; such holdings would have the effect of raising the price of raw materials to potential entrants, thus limiting their ability to compete. Thus, the structure of the industry (i.e., the number of firms) would be the result of firm conduct rather than a determinant of conduct. More generally, industry structure, rather than being a natural result of underlying supply and demand characteristics, is seen now to be endogenous to firm (and regulatory) behavior.[7]

The improved understanding of the relationship between industry structure and firm conduct has prompted a notable shift in modern industrial organization analysis away from empirical studies and toward the use of formal theoretical models. Analysts seek through such models to identify the potential conditions under which entry-deterrence strategies of incumbent firms are credible and thus have the effect of discouraging or limiting potential entrants (e.g., Dixit, 1980; Gilbert, 1986). The lessons and experience gained through such modelling exercises have given rise to a certain skepticism over arguments based on the classic SCP models that point to a concentrated structure as strong evidence of non-competitive conduct or inefficient performance without setting forth a credible model for how this structure is resistant to potential entrants. For our purposes in studying trade relations in the wild rubber sector, this critique becomes particularly important because it underscores the need for a careful delineation of what strategies might be used by incumbent firms to create the opportunity for monopolistic or oligopolistic behavior.

The second major area of revision to the classic SCP model—and one of perhaps even greater salience to our study of wild rubber—has been in the

---

[7] The endogeneity argument can be usefully demonstrated by the counterfactual scenario of the aluminum industry example. If the early entrants had not behaved strategically to acquire scarce bauxite and hydroelectric reserves, or if government regulation had taxed them sufficiently to make holding the reserves off the market too costly, then the structure of the industry would have been different because later entrants would have been able to obtain competitive raw material sources for production.

area of transaction costs (see Coase, 1937; Williamson, 1985) and the economics of information (see Stiglitz, 1986). Recent attention in industrial organization models to the potential costs associated with searching out, negotiating, executing, and monitoring transactions in an industry or activity (i.e., "transaction costs") raises a distinct set of questions about industrial organization. Such questions tend to focus on the types of contractual relations or vertical linkages among participants that might be necessary to reduce these costs and allow efficient transactions to develop. Classic SCP analyses tended to view most complex contractual forms, especially those involving durable relations between contracting parties, as a form of monopoly relations meant to limit competition. In contrast, the transaction cost approach attempts to understand the degree to which these contractual relations may be a means of overcoming incentive problems or of minimizing information costs associated with exchanges that are difficult to arrange, execute, or monitor. Monopoly intentions are not ruled out of the analysis of markets where complex contractual forms arise, but neither are they given exclusive attention. Instead, issues of proper incentive design and ensuring that agents behave in certain ways also become credible factors in explaining the existence of complex contractual forms.

Transaction costs are particularly important in certain, though certainly not all, industries and sectors. Attention to such costs becomes especially important where risk is prevalent or pervasive (e.g. price and output fluctuations) and not easily observed by all parties; where exchanges and effort are not readily monitored to ensure that agents are behaving as promised; or where the value of assets or resources deployed in an activity are, to some extent, specific to that sector (e.g., sunk investments). In such instances, complex contracts or joint governance structures (e.g., a firm) may be necessary for agents involved in distinct stages of a market process to assure that resources are allocated efficiently. To illustrate the potential role of transaction costs in creating the need for complex contracts or governance structures, we offer two examples, one from commercial agriculture and the other once more from the aluminum industry. In agriculture, contract farming is a rather common phenomenon whereby a farmer produces a certain crop under contract to a processor or seller; the contractor provides the farmer with working capital, such as seeds, chemical inputs, and even cash loans to produce the crop. The farmer delivers the produce at a price agreed upon in the contract and clears his responsibilities with the contractor. In a world of perfect information (i.e., one with no transaction costs and uncertainty), the farmer could obtain a loan directly from credit markets rather than the processor and, at the end of the season, sell the produce on the open market. Unfettered access to credit thus would obviate the need for a contract with the processor. Alternatively, the processor or seller could either buy the product on the open market or integrate backwards into production, set up a corporate farm, hire farm labor, and undertake production within the firm.

Agricultural production, however, proceeds in a world of imperfect information where risk and uncertainty can be considerable. Yields are contingent on a wide variety of environmental and human factors. A poor harvest, for instance, can be attributed to bad seed, adverse weather, pests, theft, poor crop management or low labor effort, and only the farmer can known with certainty the specific and true cause. Given such uncertainty, farm owners will monitor the effort of agricultural workers hired on a wage basis. For similar reasons, a poor farmer may have difficulty obtaining a loan from credit markets. With a weak collateral base (i.e., small land holding and few assets), the lender may not be willing to loan for the production because of the potentially high costs involved in monitoring the farmer's use of the loan and/or the repayment. In other words, the uncertainty involved in agricultural production and the information problems inherent in monitoring certain types of transactions, such as loans and labor hiring, create the potential for serious agency problems. This outcome, in turn, creates the incentive for parties to seek alternative arrangements—such as contracts—that make monitoring less necessary, or reduce the costs of transactions.

Contract farming can be viewed as an arrangement that delivers much needed working capital to poor farmers while enabling the processor to obtain a regular supply of the produce without having to hire labor directly (and hence incur the associated monitoring costs). Under contract, the farmer has the incentive to deliver as much produce as possible; however, should the farmer see that the open market price is higher than the contract price, he would surely be tempted to divert at least some of the produce away from the contracting agent. For this reason, contracts usually stipulate an expected volume of output per land unit area which will be delivered to the processor, and occasional visits during cultivation help the processor to know whether the growing conditions and farm management are likely to result in a harvest at the target levels. Contract farming thus can be considered as a means to overcome major transaction costs that arise over credit and labor (though, of course, such arrangements face their own, presumably less forbidding transaction costs). It is important to note here that this view of contract farming does not preclude the possibility that the farmer holds a weak bargaining position vis-a-vis the processor when it comes to the contract; the potential for this sort of power to arise would depend on the possible alliances among growers, competition among processors to secure growers, and the like.

The second example explores the need for alternative governance structures that can arise with transaction costs. The example is drawn from Stuckey's (1983) analysis of vertical integration of aluminum companies. In studying the first step in the making of aluminum—the processing of bauxite into alumina—Stuckey observes that the necessary processing facilities have a large minimum efficient scale (i.e., scale of lowest unit cost) and that such facilities can handle only a certain type of bauxite (i.e., one with particular physical and chemical characteristics, from a specific locale). Because investment in the alumina facility is thus both substantial and highly specific, the supply of the particular type of bauxite must be secure to avoid the risk of idle capacity or shut down. To ensure a steady supply of the specific type of bauxite needed, the firm could purchase the bauxite reserve and internalize the transaction. Alternatively, the firm could enter into a long-term contract with the bauxite owner. The problem with the latter strategy is that contract terms could not be made fail-safe (i.e., absolutely guaranteed). The high sunk cost of the alumina facility and the transaction costs associated with ensuring a stable, long-term contract combine to make vertical integration of bauxite and alumina refining an attractive governance structure for the firm, even though the bauxite mine typically lies at great distance from the alumina facility.

TRANSACTION COSTS AND EXTRACTIVE INDUSTRIES

Transaction costs issues are likely to be especially important in the study of resource extractive industries for the reasons implied in the two examples just described. First, extractive industries frequently involve labor activities that are inherently difficult to monitor, especially if the extraction is undertaken in a decentralized or geographically dispersed manner. In extractive activities, labor effort can be very costly to assess by direct observation, and like agriculture, labor effort remains fundamentally contingent upon risk factors and uncertainty (e.g., health, motivation, local conditions) that therefore indirectly affect output, but which also are costly to monitor. As a result, a prospective labor contractor would face potentially serious problems: specifically, how to ensure that engaged labor would work hard, and how to ensure that all of the product extracted by the laborer would be delivered to the contractor. Second, extractive industries can involve large sunk investments, i.e., where capital cannot be readily recovered, either in equipment and infrastructure to establish an extractive facility (e.g., open pit mine and railway) or for provisioning of personnel working in remote areas (e.g., in logging or gold panning). Yet, the very remoteness of extractive activity introduces information problems which, in turn, will tend to create the need for contractual relations or governance structures that reduce the uncertainty around returns to the fixed investment. Third, the risks involved in resource extraction, and the monitoring problems mentioned above, will tend to limit the extent of arms-length credit made available to those directly involved in extractive activity. At the same time, the risks and needs for provisioning may actually increase the demand for credit among industry participants. As we shall see later, specific contracts emerge to deal with such problems. Clearly, the role of transaction costs and risk are of special importance to understanding the organization of extractive industries.

The transaction costs approach, however promising, has yet to be fully integrated into modified SCP models or the literature on industrial organization. The reasons are many. Transaction costs are not prevalent in all industries, and the importance of such costs in those where they are substantial is often underestimated. More importantly, such integration represents a formidable conceptual challenge. To explain firm conduct and industry structure, where transaction costs are significant, essentially requires that another layer of economic activity be added and accounted for within the SCP model. As the reader may have noted from our examples above, the contractual forms and governance structures that emerge to overcome transaction costs may (or may not) lend themselves to competitive market outcomes; whether they do, depends upon not only those conditions typically considered in industrial organization analysis but also the effect of institutional arrangements themselves. To include consideration of the endogenous formation of contracts or other institutions that emerge to govern transactions would render the analysis considerably more complex. Such arrangements may (or may not) provide industry participants with strategic opportunities to affect industry structure (via entry) and hence the nature of competition. Not surprisingly, therefore, integration of the transaction cost approach and modified SCP models is by no means complete in the industrial organization literature. Our purpose here is not to achieve such integration, but rather to use insights that flow from such approaches to better understand the wild rubber industry.

Toward a Model of the Organization of Extractive Industries

The classic SCP model provides a useful starting point for our analysis of the Amazonian wild rubber industry as well as other extractive activities because the model rightly points our attention to the importance of the basic characteristics of an industry. Such characteristics can be defined by whether they lie on the demand side or the supply side of the extractive industry, and they can be elicited through focused inquiry around on a number of key questions. On the demand side, the following issues become important: What is the nature of demand for this extractive product? Is the product essential in the sense that there are few, near substitutes? Is demand for the product strong and growing? How diverse are the demand sources? On the supply side, analysis would focus on a different set of questions: What are the key factors of production? How abundant, in relative and absolute terms, are such factors? Who controls the factors of production? What technologies are used to transform these factors into the product? What form and shape do the resulting cost curves take? Do obvious barriers to entry exist that can be identified in these supply conditions? Can existing firms or economic agents exploit such barriers?

In addition to the classic demand and supply side features of the extractive industry, the analysis of the organization of the industry must also consider the problems of risk and transaction costs. More specifically, the following issues become relevant: What types of risks are there at various stages of resource extraction, transport, and trade? What types of activities and exchanges generate major transaction costs? What kinds of contractual forms or governance structures might be used to improve the incentives facing agents and reduce costs? From consideration of such issues, with the demand and supply features of the industry, attention can turn to the conduct of industry participants. Here, the analysis would seek answers to questions such as: How do the supply, demand, risk, and transaction cost characteristics shape the options for conduct available to various agents? How do endowments, such as wealth, timing of entry to the industry, knowledge, and other factors, affect their ability to act strategically or to obtain better terms from certain transactions? From the resulting view of firm conduct, it is then possible to explain the basic structure of the extractive industry, the competitive behavior of its participants, and the performance outcomes.

Chapter 3

RUBBER INDUSTRY CHARACTERISTICS: MARKETS, EXTRACTION AND TRADE

Introduction

Throughout the Amazon Rubber Boom, virtually all of the rubber extracted from the forests of the basin was exported to world markets. For this reason, we begin our account of the basic characteristics of the wild rubber industry with an examination of the international rubber trade during the late 19th and early 20th centuries. Of particular interest are the nature of the emergent trade in wild rubber and Amazonia's position in world rubber markets. Having established the market context for the reader, we turn then to look inside the wild rubber industry. We describe the various types of rubber extracted, the basic technologies involved in extraction and transport—including the necessary productive inputs—and the risks and transaction costs involved in those operations. This account of the markets, extractive process and the trade in wild rubber sets the stage for a discussion of the conduct of industry participants in Chapter Four.

The International Wild Rubber Trade (1860–1920)

Two issues are paramount in understanding the nature of the international rubber trade during the Rubber Boom: the relative position of Amazonia in the world provision of crude rubber; and, the degree of competitiveness of the international rubber market. In general terms, when a region is a primary source of a raw material and alternative sources are of higher cost, take time to develop, or do not exist, opportunities are created for the local capture of economic rents from extraction of the product. If foreign companies hold substantial market power, then such firms potentially can limit market prices as well as returns at the extractive end of the industry. Alternatively, in a competitive international trading system, market prices for the resource at the point of extraction are more likely to reflect the true scarcity value of the product. Thus, if the region is the prime source of the raw material and the international market is competitive, then firms within the region should be able to capture a significant portion of the high returns associated with raw material extraction.

AMAZONIA AS PRIMARY WORLD SUPPLIER OF WILD RUBBER

From the rise of wild rubber markets in the 1860s to the early 1900s, Amazonia was the primary source of the finest grades of crude rubber in the world. In the early years, Amazonia was virtually the sole supplier of wild rubber; even during the 1890s and the early 1900s, when African and Central American production had expanded significantly, the basin contributed approximately 50 percent (sometimes more) of world production (see Figure 2). Within Amazonia, Brazil was clearly the dominant producer, supplying between 80 and 90 percent of total basin production in any given year, followed by Peru and Bolivia (roughly 5–10 percent each) (see Table 1). Colombia, Ecuador, and Venezuela were incidental producers, together supplying less than a few percent of total Amazonian output (see Iribertegui, 1987:168–170).

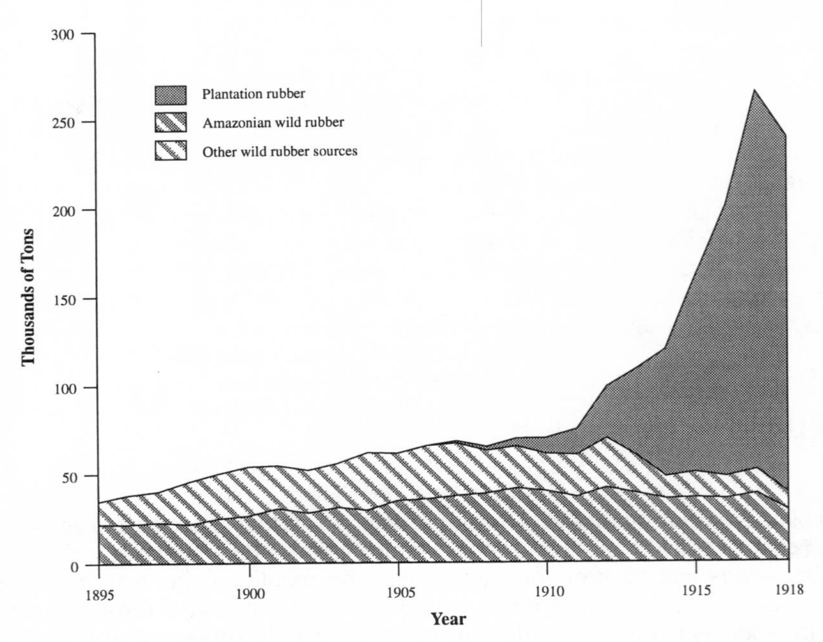

Source: Data from *India Rubber World*, vol. 47, no. 3 (Jan. 1, 1913):197, and *India Rubber World,* vol. 60, no. 6 (Sept. 1, 1919):725.

Figure 2. World Rubber Production, 1895–1918

TABLE 1. WILD RUBBER EXPORTS DURING THE RUBBER BOOM FROM LEADING AMAZONIAN PRODUCERS, 1890-1910

Exports of Amazonian Wild Rubber: Hevea and Caucho (millions kg)

YEAR	BRAZIL	PERU	BOLIVIA
1890	15.4	1.1	0.3
1891	16.7	1.5	0.4
1892	18.3	1.5	0.4
1893	18.3	1.2	0.4
1894	18.5	1.3	0.6
1895	19.0	1.0	0.8
1896	19.5	1.2	1.1
1897	20.7	1.9	1.7
1898	19.8	1.8	3.2
1899	22.9	1.4	2.1
1900	23.7	1.2	3.5
1901	27.9	1.2	3.5
1902	27.1	1.5	1.9
1903	29.1	1.3	1.3
1904	27.1	2.2	1.6
1905	31.9	2.5	1.7
1906	31.4	2.8	1.9
1907	34.5	5.1	1.8
1908	34.3	5.0	2.6
1909	34.7	4.4	3.1
1910	34.3	2.3	3.1

Sources: Brazil (LeCointe, 1922:I:434); Peru, 1890-1909 (Maúrtua, 1911:27); Peru, 1910, Port of Iquitos only (Fuller, 1912:922); and Bolivia (Ballivián and Pinilla, 1912:248-49).

The superior quality of Amazon rubber was widely recognized in world markets. Indeed, the industry's prime grade—Pará upriver, fine old rubber—was named after the region near the mouth of the Amazon River where extraction first boomed; it was the industry standard throughout the boom when as many as 40 different varieties of wild rubber were known in the market. Buyers would often pay premiums of up to 30 percent for Pará fine rubber over alternative supplies from Africa and Central America (see Table 2). This quality advantage was an important part of the Amazon's dominant position in world crude rubber supply, as was the relative abundance of rubber trees in the basin.

TABLE 2. NEW YORK PRICE QUOTATIONS FOR TEN TYPES OF WILD RUBBER (OCTOBER, 1890–1910)

	Price (cents/lb)			
	1890	1900	1905	1910
Amazon				
Pará, fine	73-74	92-98	118-133	121-143
Pará, coarse	50-51	52-73	68-90	73-103
Caucho, ball	58	65-66	85-86	100-101
Central and South America				
Esmeralda, sausage	55-56	63-64	82-83	91-92
Guayaquil, strip	44-48	56-57	70-71	—
Nicaragua, scrap	55	63-64	81-82	90-91
Africa				
Sierra Leone	45-50	68-70	100-101	119-120
Benguella	52-53	62-63	79-80	88-89
Congo, ball	44-45	57-58	110-111	110-111
Asia				
Borneo	35-50	37-48	44-45	—

Sources: *India Rubber World and Electrical Trades Review*, vol. 3, no. 2 (Nov. 15, 1890):59 for "most recent quotation" (i.e., end of October); *India Rubber World*, vol. 23, no. 2 (Nov. 1, 1900):60 for October 31st; *India Rubber World*, vol. 33, no. 2 (Nov. 1, 1905):64 for October 31st; and, *India Rubber World*, vol. 43, no. 2 (Nov. 1, 1910):71 for October 28th.

INTERNATIONAL MARKET STRUCTURE

During the boom, the bulk of wild rubber production from the Amazon was exported by a small number of foreign-owned export houses. In 1904, for example, the top five exporters from Belém (known then as Pará) and Manaus accounted for 80 percent of rubber shipped to international markets from the Brazilian Amazon (see *India Rubber World*, vol. 31, no. 5, Feb. 1, 1905, p. 180). Similarly high export concentration levels existed throughout much of the boom period in Brazil, Peru, and Bolivia, and, at first glance, would seem to support the contention made by some scholars (e.g., Flores Marín, 1987) regarding foreign monopolization of the Amazon rubber trade. However, the existence of high concentration levels alone is not sufficient to demonstrate oligopoly control; one must also show the presence of significant barriers that prevent other firms from entering the market to capture the higher than normal returns allowed by market power. Alternatively, high concentration levels can be consistent with competition, if the contestability of the market limits the potential for oligopolistic behavior (Baumol, Panzer, and Willig, 1982).

Two related arguments suggest that the international rubber market was highly competitive. First, the market environment of the international rubber trade is consistent with the common features of a competitive market. Wild rubber was traded freely in major markets of Europe and the U.S. as well as in the port-cities of Amazonia. Rubber trade in Amazonian ports was vibrant. Export houses depended not only upon contracted suppliers of rubber; they also purchased their rubber at auctions and sales from local traders. Prevailing prices were openly quoted in both export and import markets, and rubber was graded by location and quality in the dock-side export houses for shipment overseas.

International crude rubber markets were also transparent. Industry information was abundant, with trade journals, such as *India Rubber World* (published in New York) and *India Rubber Journal* (published in London), reporting on shipments by different trading houses from port of departure and by arrivals, on price quotations for different grades of rubber in the major exporting and importing markets, and on the reshipment of primary rubber from one importing market to another. Antwerp, Havre, Liverpool, and New York were the four major international exchanges for rubber, and the trade journals provided timely information on the activities and new investments reported from all four markets.

Second, compelling evidence of the industry's competitiveness comes from the absence of observable entry barriers for new participants at any stage of the industry. Export activities were easily staged: export houses obtained rubber from local traders in both contractual and spot market transactions, verified grades, and then shipped and sold the rubber, none of which posed obvious barriers to entry. Grading was a process of standardization that, in effect, precluded private strategies of differentiation of the primary material and enabled competition by setting general quality standards for the industry. And shipping was then, as it is now, the essence of a contestable activity. New entrants could enter a shipping route when profits were high, then move their boat to another market when profits fell. Moreover, the economies of scale involved in transporting a greater volume of product over long distances meant that shipping had naturally higher concentration levels than other stages of the industry. As rubber constituted only a minor proportion of international shipping by volume, and this period involved no massive naval disruptions, there was no shortage of shipping capacity to prevent competition at this level of the industry. The manner by which wild rubber was marketed, given grading, public auctions, open trade channels in import markets, and significant transshipments of rubber across these major markets is also consistent with open entry into trade.

The only remaining stage where potential barriers could have impeded entry to the trade is in the acquisition of wild rubber, either from tappers or traders, for shipment and sale overseas; however, few effective barriers to entry can be detected at this stage. Opportunities to enter the trade by securing rubber were virtually unlimited. Not only was rubber procured from several different regions within the Amazon basin (and elsewhere in the world) but sourcing of wild rubber everywhere was an extensive and highly decentralized activity. As argued below, the decentralized nature of rubber collection made the work costly to monitor and control, and even though lines of credit extended out from the relatively few, city-based exporters to the tens of thousands of tappers in the forest by trading houses and traders, virgin rubber trees remained to be tapped by newcomers. Itinerant traders flourished, backed by urban merchants who contracted for the purchase of rubber on the river.

In the major port-cities along the Amazon, hundreds of small trading houses dealt in rubber, none of which controlled more than a fraction of the exportable rubber. At Manaus during 1909, for example, 45 firms are reported to have received at least 100,000 kg of rubber, accounting for 83.3 percent of all rubber entering the port; the remainder was received by over 900 small firms (see Loureiro, 1986:222, 230–235). Pearson (1911:41) observes that from Belém, a small steamer in a single trip from the Islands district may carry rubber from 200–300 shippers destined for 75 or 80 different rubber receivers. Though many traders had contracts with the larger export houses, they also sold rubber in local exchanges. Moreover, trade contracts could be renegotiated or bought out; often trading houses would have their debt with one exporter bought by another to become a supplier for a different exporter.

In sum, Amazonian rubber collection, trade and marketing was too decentralized, too extensive and too open to entry for much market power to be acquired by a small group of participants. Of the impediments to a world rubber monopoly, the editor of *India Rubber World* wrote in 1901,

> No fear need be felt that the supply of rubber will ever be monopolized. Could such a thing be done, the manufacturer, first, and the consumer of rubber goods, in the end, might be made to suffer exactions hard to bear ... A complete monopoly of rubber ... would mean a great general trading company, constantly liable to competition from new sources, in respect not only to rubber, but to all the other commodities ... Rubber ... is obtained throughout a wide belt, extending around the world, but mostly in regions remote from civilized centers...; it comes in driblets to innumerable initial markets, from millions of gatherers, whose labor practically is beyond control. Moreover, if all the forests now yielding rubber, and all the rubber gatherers at work, and all the houses trading in rubber in America, Europe, Asia, and Africa, were brought under a single control, the possibility would exist of new forests being explored, new workers found, and new trading houses opened, every one of which facts would tend to the overthrow of the monopoly. (Pearson, 1901:135)

Indeed, the Brazilian trading house of Vianni repeatedly attempted, without success, to corner the Belém rubber exchange over more than a decade in the 1870s and 1880s, when the rubber trade was still relatively young (Weinstein, 1983b:139–155). The government of Brazil attempted later to valorize rubber by purchasing rubber on the open market in Belém and holding back supply, again without much success.

Numerous discussions of new investment schemes and projects in the trade journals coupled with the frequent entry and exit of export houses during the boom are also consistent with an actively competitive industry. Wild rubber thus contrasts sharply with other natural resources, such as bauxite and diamonds, whose scarce, concentrated, and appropriable prime reserves made possible the formation of monopolies or trusts through control over the raw material reserves.[8] In the case of wild rubber, Amazonia retained its lucrative natural monopoly on Pará fine rubber for some 50 years, and both the world market and the international industry which developed around external demand for crude rubber were highly competitive.

The prime position of Amazon wild rubber in a competitive world market created the conditions for an unprecedented economic boom based on rubber extraction and the potential for significant capture of resource rents by local industry participants. The boom attracted numerous investment ventures in extraction as well as trade from Europe and the United States as well as from domestic firms. Tens of thousands of immigrants were drawn into the forests of Amazonia to work in rubber extraction, transport, and related activities. A clearer appreciation of the economic development logic of the boom thus depends more on understanding how the returns and risks associated with the rubber industry were distributed across participants and regions within Amazonia than on issues of foreign control or surplus drainage. Any analysis of the returns and risks associated with rubber collection must consider in more depth the organization and microeconomics of the industry. To that task, we now turn.

An Overview of Rubber Extraction and Trade

Several basic characteristics of the natural endowment of wild rubber in Amazonia strongly shaped the organization of the rubber industry. Of singular importance was the distinction between *hevea* and *caucho*, the two principal types of rubber gathered from the forest. Hevea rubber was obtained from species of the endemic hevea tree (primarily, *Hevea brasiliensis*, *H. benthamiana*). The most valuable rubber latex was extracted from *H. brasiliensis* which grew at relatively low densities (e.g., 28 trees/km^2) (Weinstein, 1983b:17) primarily south of the Amazon River, in a band from Belém to the Ucayali River in Peru (see Figure 3). Although the

---

8 The determinants of monopoly and strategic behavior by firms in scarce resource industries are explored in Barham (1994), Barham *et al.* (1994), and Barham (1991). Relative scarcity of the material and a high spatial concentration distribution of the material are necessary conditions if a few firms are going to obtain market power by controlling access to the scarce resource. In the case of wild rubber, the scattered distribution of rubber trees in the rain forest made this type of strategy impossible.

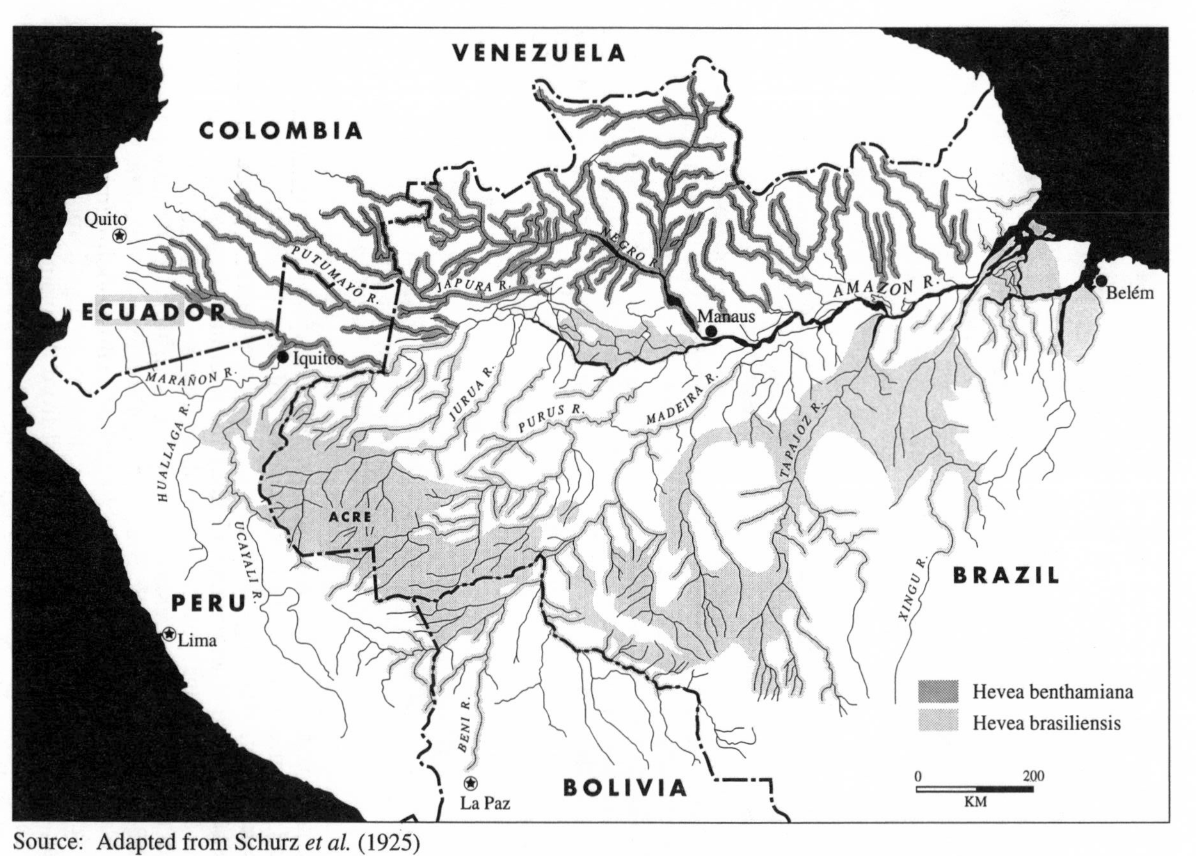

Source: Adapted from Schurz *et al.* (1925)

Figure 3. Distribution of Hevea Wild Rubber Trees in Amazonia

individual hevea tree produced only a small quantity of latex upon tapping (i.e., 25–50 gms/tree/day), the latex could be extracted every other day for many years and when cured over a smoky fire became the finest rubber in the trade. In contrast to hevea, caucho was not suited to regular tapping. The less valuable rubber was extracted through the felling and bleeding of species of the castilloa tree (*C. elastica*; *C. ulei*) found most commonly at low densities or in small stands deep within the upland forests of Upper Amazonia, in an arc from Colombia through Ecuador and Peru to Bolivia.

Hevea latex, from which the highly appreciated Pará fine rubber was derived, dominated the Amazon rubber trade. Hevea rubber exports from the Brazilian Amazon during 1890–1909 exceeded caucho exports by a ratio of approximately 5–10:1 (derived from Pearson, 1911:215). Peru, which contributed less than 10 percent of total basin output of rubber, exported a significant proportion of rubber as caucho. Prior to 1897, caucho dominated Peruvian production, and thereafter hevea made up roughly 50–70 percent of annual rubber exports from that country (derived from Maúrtua, 1911:27).

The organization of wild rubber extraction varied according to the type of rubber sought. Where hevea trees were present, along the southern tributaries and in Acre, rubber trails (*estradas*) were opened in the forest to connect the dispersed trees and allow for frequent visitation. Collectors of hevea rubber (*seringueiros*) typically each worked two estradas (Figure 4). Each estrada contained some 80–150 hevea trees, was tapped on alternate days, and covered an area of 3–5 km. A seasonal and solitary activity, tapping was conducted when floodwaters receded and rains were less frequent. The rubber season varied in duration and timing during the year, depending on the location in the basin, but usually lasted less than six months.

The fixed capital requirements for tapping, once the estradas were established, were minimal and tools were limited to a small cutting axe, cups for collecting latex, and a pail for carrying the latex to the tapper's hut (Photo 1). Processing comprised the daily curing of hevea latex that over time would form large oval rubber balls, each weighing 10 to 100 kilograms (Photo 2). Seasonal yields varied considerably among tappers and locations but a typical tapper could have processed some 250–350 kilos of rubber in 100 days of work. During the off-season some tappers migrated to nearby settlements while others remained along the river until hevea collecting could begin anew. At the peak of the Rubber Boom (ca. 1910), an estimated 131,000–149,000 men were tapping hevea from some 21.4 million trees on 24,000–27,000 estates in the Brazilian Amazon (Santos, 1980:66, 83, 84).

Caucho was collected by highly mobile teams, often numbering 15–50 men (*caucheros*), that would comb extensive areas of upland interfluvial forest, venturing many kilometers inland from the main rivers in search of caucho. Once caucho trees were located they were marked and then fallen, drained of latex, and abandoned (Photo 3). Latex was coagulated on site using soap or natural products and each tree yielded some 20–30 kg of caucho (LaRue, 1926:60). A skilled cauchero would collect 500–1000 kg of

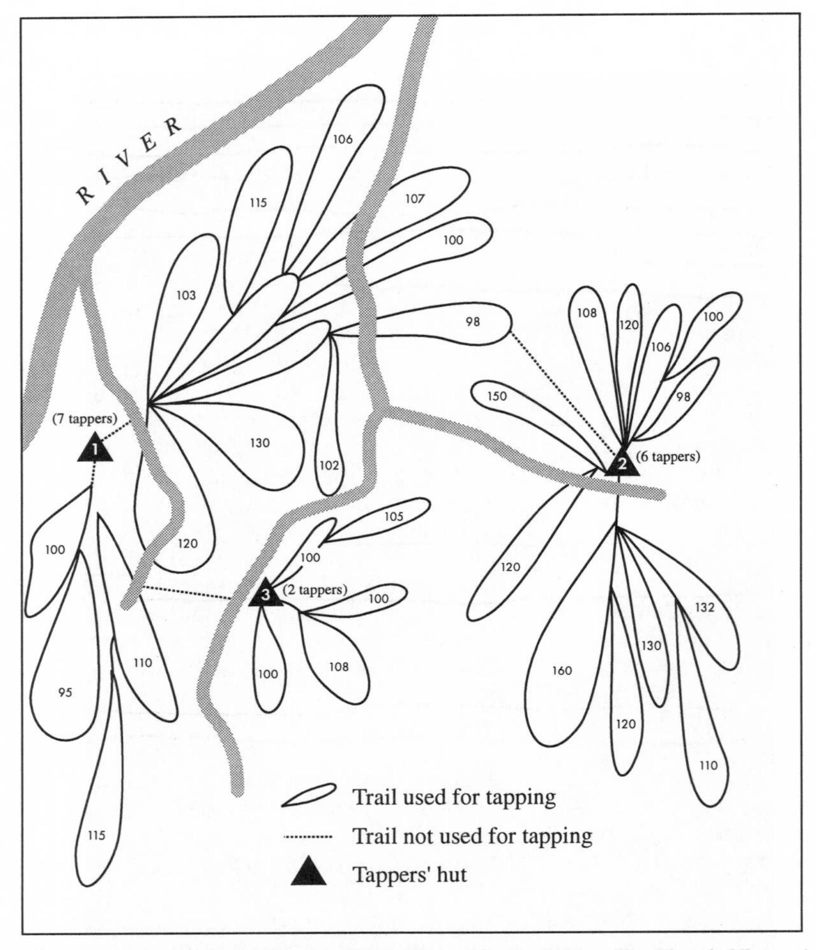

Source: Adapted from Pearson (1911:13) and *India Rubber World*, vol. 27, no. 1 (Oct. 1, 1902):15.

Figure 4. Plan of Peruvian Rubber Estate, Showing Estradas, Number of Trees per Estrada, Location of Tappers' Huts, and Number of Tappers per Hut

caucho over a period of at least six months (see Fuentes, 1908:I:214; Woodroffe, 1914:105), though harvests of 2000–3750 kg were known (see LaRue, 1926:60; Yungjohann, 1989:61). Data from the Putumayo and other regions where native peoples worked caucho under abusive overseers suggest much lower yields, between 40–120 kg/collector/season (see U.S. Congress, 1913:68, 86). Unlike hevea, caucho could be worked in any season and thus some hevea tappers extracted caucho during the off-season in forays to the more remote, inland forests.

Photo 1. Hevea Rubber Tapper, ca. 1910
(from Lange, 1912:151)

Photo 2. Curing and Branding of Hevea Rubber
(from *India Rubber World*, vol. 42, no. 4, July 1, 1910:343)

Photo 3. Extraction of Latex from Caucho Rubber Tree
(from Pearson, 1911:155)

The trade in extracted rubber was strongly influenced by the great distances that separated the gatherers from buyers in the wild rubber industry. Provisions were advanced by on credit by urban-based trading houses and river traders to enable rubber workers along thousands miles of rivers and streams of the basin to gather rubber intensively without the need to devote valuable time to agriculture and other subsistence activities. In return, relatively small quantities of rubber from throughout the basin were funneled by traders who braved rapids, shifting sand bars, low water, and theft to deliver the rubber to the major Amazon port-cities where the product could be exported overseas in large batches on ocean-going steamships (Photos 4–7).

Whereas the journey from the export houses of Belém at the mouth of the Amazon river to rubber exchanges in Europe or the United States is indeed appreciated as long in both distance and time, it was the voyage between the rubber fields and Belém that reflects the true remoteness and degree of geographical separation between the gatherer and the eventual industrial or urban consumer. A return trip, for example, from Belém to the rubber fields of the Beni River would entail 270 days, barring misfortune (Pearson, 1911:121), or the equivalent of 13 return voyages from Belém to New York.

Near the port-cities, independent hevea tappers and patrons could trade directly with trade and export houses whereas those working the distant reaches depended on intermediary traders who advanced credit and expected rubber in return. Itinerant merchant traders operated outside the traditional *aviamento* (provisionment) system along the main stem of the Amazon River and its tributaries, tempting patrons and tappers alike with goods or cash for rubber.

Risks and Transaction Costs

Industry participants faced a variety of serious risks that continually threatened the profitability and security of the rubber trade; the most salient to industry organization were the loss of product, labor time and workers, as well as fluctuations in prices and exchange rates. The risk of product loss was attributable to several factors. One was the treacherous nature of river travel. The Madeira River rapids—one of the worst sites in the region—reportedly swallowed up between 5–25 percent of the rubber shipped and foreign companies refused to insure boats that tempted fate in the rapids (see Fifer, 1970:130; *India Rubber World*, vol. 24, no. 5, August 1, 1901, p. 327; and *U.S. Consular and Trade Reports*, vol. 66, no. 248, May, 1901, p. 148). Product loss also occurred in rainy weather which spoiled the fine rubber latex, with shrinkage in transit of up to 16 percent (Schurz *et al.*, 1925:29–30), and because of impurities introduced by nature or by the sly tapper who knew that detection might not occur until the rubber was hundreds or thousands of miles away (Pearson, 1911:38–42).

Photo 4. River Front Trader's Store, Peruvian Amazon, ca. 1910
(from Lange, 1912:21)

Photo 5. Small River Launches, Yavari River, 1910
(from Lange, 1912:45)

Photo 6. River Steamer at Port of Villa Rio Branco, Acre River, ca. 1905
Property of the Companhia do Amazonas

(from Falcao, 1907)

Photo 7. Flotilla of Aviador and Itinerant River Trading Boats at Villa Río Branco, Acre River, ca. 1905

Losses of both workers and labor time in the process of rubber extraction were substantial. From first-hand accounts, overall tapper attrition rates due to death, illness, and desertion, especially in the Upper Amazon appear to have been very high indeed. A report from U.S. Consul Kenneday at Belém noted that of 100 workers recruited and sent to the rubber fields, 75 would die, desert, or leave because of illness (*U.S. Consular Reports*, vol. 59, no. 220, January, 1899, p. 70). In one specific case, 300 tappers sent upriver from Belém resulted in a net increase at the end of the year of only 5 tappers to the upriver estate, (*India Rubber World*, vol. 23, no. 2, Nov. 1, 1900, p. 34). Of 350 workers engaged at Belém to work the Madeira-Marmore railway, only 65 reached Pto. Velho (Santos, 1980:96). Typically, a patron in the Upper Brazilian Amazon could expect to lose 5 out of 25 tappers to multiple causes shortly after arrival on the estate (Woodroffe, 1914:221–22).

Other reports speak of high tapper mortality.[9] Death rates of 25–30 percent (and as high as 50 percent) have been suggested for rubber workers in Bolivia during the boom (Fifer, 1970:130). A common rule of thumb used by rubber estate owners when gauging labor requirements was to engage 80 men where the work could be done by 50 able bodies (Schidrowitz, 1911:26). Obviously, from the perspective of the patron or the trader, this loss of life also meant a loss of product and capital, and such risk would have to be factored into investment calculations. The expected loss in labor time also have been considered in such calculations. As a rule, rubber workers managed to tap rubber on only about 100 days of the six month season.[10] Tapping was cut short by rain, accidents and illness with malaria, yellow fever, and beri-beri, often costing workers weeks of the tapping season—if not their very lives. Although no systematic study of mortality rates of rubber tappers has been done, high death rates appear to be due more to disease and accidents than to violence. Tappers undoubtedly would have considered the risk to life and limb in setting out for the rubber fields.

The third risk in the rubber industry was associated with the wide variations in rubber prices as well as exchange rates during the boom. Trade journals, such as the *India Rubber World*, ran regular articles attempting to explain the major price swings of rubber and why speculation was probably a minor force relative to the supply shocks that wild rubber extraction experienced. The very inelasticity of wild rubber extraction supply (and demand) discussed further below made the likelihood of large price swings rather high. In addition, however, Brazilian exchange rates were particularly volatile during the rise of Republicanism in the 1890s. Because rubber extraction was financed by debt-merchandise contracts, creditors ran the risk

---

9 See Lange (1912:89–90), Woodroffe (1914:221–222), LeCointe (1922:I:101,106), *India Rubber World* (vol. 4, no. 2, May 1, 1891, p. 209; vol. 23, no. 1, Nov. 1, 1900, p. 34), and Yungjohann (1989:45,49,61).

10 For reports of the number of days actually worked during the tapping season, see Jumelle (1903:86), Fuentes (1908:I:212), Lange (1911:34), and Schurz *et al.* (1925:13).

of receiving poor prices for rubber delivered on forwarded goods that inflation had also made more expensive to purchase for the next round of rubber gathering. Local currency devaluation also could make imported provisions more expensive; if accompanied by falling rubber prices, this price-cost squeeze could severely pinch creditors.

The nature and magnitude of risk involved in rubber extraction would suggest that risk premia were built into the returns to key factors of production, especially labor and capital which, as discussed in the next chapter, were both mobile and scarce. Such premia can be expected to have varied substantially depending on the location of rubber gathering, with the downriver environments being, on average, less risky for both transport and perhaps labor health. Furthermore, premia surely varied with the bargaining position of different rubber industry participants, with the more desperate inmigrants and the less market-based Amerindian peoples most likely receiving much less of a risk premium than other participants.

In addition to high risks, the Amazon wild rubber industry also faced substantial transaction costs. As discussed in Chapter Two, transaction costs are incurred as participants search out each other, negotiate an exchange, monitor its execution, and/or enforce the conditions of the exchange. The more searching, negotiating, monitoring, and enforcing required to transact an exchange (e.g., rubber for credit), the higher the transaction costs.[11] Several factors contributed to high transaction costs in the process of rubber extraction, including the extensive and risky nature of extractive activity, the fundamental scarcity of labor which lowered the costs to workers of not cooperating with creditors (i.e., patrons, traders, or trading houses), and the mobility of labor and capital.

Hevea extraction on rubber estates exemplifies the burden of transaction costs. The scarcity of labor in the region meant significant search and transport costs to bring workers long distances to the estate. Coupled with the subsequent risks of loss of labor to death, illness, or exit, the acquisition of a steady labor force was a costly endeavor. Moreover, rubber tappers worked in isolation, gathering a small volume of latex to be processed at their hut at the end of each day, which along with the high cost of labor made frequent monitoring (e.g., daily) of their effort or output prohibitively expensive. Tappers were well placed not to deliver all of the collected rubber to the creditor, selling instead to itinerant traders or shipping the rubber themselves; the apparently low output could be attributed to illness, rain or theft. The creditor—unable to afford close monitoring of the tappers daily activities—would have no ready basis for rejecting the tapper's explanation.

---

[11] More generally, transaction costs are considered to be the costs of running an economic system "the economic equivalent of friction in physical systems." Oliver Williamson (1985) distinguishes transactions costs from direct production costs as the costs that come, in some sense, prior to and after production. Prior costs include the costs of finding the agent(s) with whom exchange or transactions will occur and then drafting, negotiating, and safeguarding an agreement with them. Costs that arrive after production include the process of correcting misalignments, the set up and maintenance costs of structures to govern disputes, and the bonding costs of securing commitments.

Even worse, labor's fundamental mobility also raised the potential for tappers to disappear with rubber earnings without paying their debt. The risk of desertion would be related, in part, to the terms offered to the tapper; presumably, the worse the terms, the more effort estate owners would have to spend on trying to constrain tapper mobility.

Trade in rubber also was subject to high transaction costs. The vast geographical area over which large investments (e.g., merchandise and future claims on rubber) were made and the nature of the business also made monitoring of traders and patrons—to ensure that they sought and delivered returns and did not abscond with either the principal or the payments—potentially very costly and problematic. Patrons and traders themselves were not above hiding a portion of collected rubber from their respective creditors and selling it to itinerant traders or merchants for cash while seeking forbearance on merchandise debts (see DeKalb, 1890:433). Assessing the veracity of claims by patrons and traders of low tapper productivity would not be straightforward. In addition, neither patrons or traders could easily verify that the rubber received from tappers was free of foreign materials introduced to increase its weight.

The combination of multiple risks and high transaction costs involved with monitoring behavior among different participants in the wild rubber trade meant that certain types of market exchanges were likely to be problematic. For example, hiring a worker to tap rubber on an estate for a wage was likely to require extensive monitoring of the output (i.e., regular visits to the tapper and his estradas) to ensure that a good effort was being exerted and that the full quantity of tapped rubber was being delivered. Similarly, if a rubber exporting house were to hire its river traders for a wage, they would also face monitoring problems on both effort and full delivery which would be hard to separate from the legitimate risks associated with river travel. In this way, conventional types of labor-capital relations were ridden with risk-transaction cost problems; such problems created the need for distinctive contractual arrangements that would help improve the incentives for participants to transact more efficiently.

Key Characteristics of the Amazon Wild Rubber Trade

As the first step in our analysis of the industrial organization of the wild rubber industry, this chapter set forth what we consider to be the influential supply and demand characteristics. Excluded from our discussion have been the basic factor endowments which are described in some depth in the following chapter. The key industry characteristics identified above are summarized as follows:

- The Amazon basin was the prime source of a highly valued raw material, distinguished by its superior quality and majority market share over several decades;

- The international trading system was highly competitive, because supply was widely dispersed, readily accessible by new entrants, and bundled together by literally hundreds of exporting firms in the main ports of the Amazon;
- The prime market position of Amazonian rubber and competitive nature of the trading system together meant that local returns for extractors of this highly valued commodity would embody a good premia for being the lead supplier of this high quality rubber;
- The main source of wild rubber was the endemic hevea rubber tree, found scattered within the rain forest, from which individual tappers, working large areas of forest, would draw latex every other day by carefully delivered hatchet cuts to the trunk, and then cure the latex into transportable balls of crude rubber.
- River traders—both regular and itinerant—who transported the crude rubber out of the rain forest and to the river port-cities, were responsible for supplying patrons and their tappers with provisions that enable them to dedicate themselves primarily to rubber tapping.
- Risks and transaction costs were substantial in this dispersed and extensive economic activity for all parties involved in rubber extraction and trade, including tappers, traders, export houses, and owners of the estates.

With these features of the wild rubber industry in mind, we now turn our attention to explaining the conduct of rubber industry participants, especially the types of contractual relations which emerged to overcome the risks and transaction costs associated with the extraction and trade of wild rubber.

Chapter 4

RUBBER INDUSTRY CONDUCT: FROM ENDOWMENTS TO RELATIONS AND RETURNS

Introduction

In this chapter, we examine the dominant forms of property held by participants in the wild rubber industry, the key labor and capital relations that emerged within the industry, and the level and distribution of financial returns realized by industry participants. We begin by introducing the notion of relative factor scarcity and by demonstrating the essential role of factor scarcity in shaping the organization of primary resource industries. We employ the factor endowment argument to structure our analysis of the wild rubber industry. Initially, we re-examine the nature of the key endowments—namely rubber trees and land, labor to tap the trees, and capital to finance extraction and trade in wild rubber—emphasizing the underlying causes of their relative scarcity. We then examine how such factors were brought together in specific relational forms for the extraction and marketing of wild rubber. Such relational forms, in turn, are suggested to have profoundly influenced the level of financial returns that flowed from the various rubber-related activities as well as the distribution of returns across industry participants, from rubber tappers to river traders.

The Relative Factor Endowment Argument

In a decentralized, market-based industry, the relative availability of the principal factors of production (i.e., land, labor, and capital) plays a fundamental role in shaping industry organization and the distribution of financial returns. In general, an industry will tend to be organized in a manner that economizes on the scarcest factors of production and such factors will bring a premium over the more abundant factors. In an agricultural region, for example, where land is abundant relative to other factors, production will tend to be organized in an extensive manner, with choices of crops and techniques of cultivation that take advantage of the abundant land and economize on other more scarce resources. Moreover, the more abundant a particular factor, the lower a relative return it is likely to command in the sector. So, if land is abundant, and labor is scarce, then there should be a premium on labor. Conversely, should land be scarce and labor be relatively abundant, then production would emphasize more intensive use of the land, absorbing plentiful labor and economizing on the limited available land. Returns generally would not favor laborers but rather land owners, especially those who used their land most efficiently. This argument has been applied elsewhere to understand the distinct paths of agricultural innovation that have been observed between developing and more developed countries (see Hayami and Ruttan, 1985).

To employ the relative factor endowment argument to understand the wild rubber industry requires one important caveat: the argument must be applied to the industry at the regional level and over the course of the boom. If used to explicate specific factor relations or predict returns at particular locales or moments, the argument could be rather misleading because property rights were often poorly defined in certain more remote reaches of the basin, markets tended to function unevenly and peculiarly, and coercive relations were evident in specific areas of the region. However, at the basin-wide scale, the argument provides a useful starting point.

Factor Endowments

The availability of three key factors—rubber trees and land, labor, and capital—fundamentally shaped the organization and evolution of the regional wild rubber industry. Each factor is considered in turn.

Rubber Trees and Land

Throughout most of the Rubber Boom, new participants could gain access to untapped rubber trees by simply moving into the more remote upriver areas of the basin and claiming an area of forest. Large areas of untapped trees remained in the basin even after the boom, particularly in the more remote regions of Brazil, such as in the Matto Grosso region (see Schurz *et al.*, 1925:39). If the capital and labor needed to establish estates could be secured, new areas then could be brought into production. Moreover, the fact that rubber extraction extended over decades into the farthest reaches of the Upper Amazon suggests that early estate owners did not have the type of preeminent strategic position that firms in other resource industries do when they hold, for example, prime mineral reserves. Moreover, even the distance-related factor of more convenient and thus potentially more valuable downriver estates was obviated during much of the era by the higher yielding rubber trees of the upriver regions (e.g., Acre).

The relative scarcity of the other two principal factors of production—labor and capital—was clearly more important.

LABOR

Labor was fundamentally scarce throughout the boom and such scarcity limited the expansion of wild rubber output from Amazonia. Labor was scarce for three basic reasons. First, the region's indigenous population was decimated by disease during the Conquest (see Denevan, 1992; Bunker, 1984:1029–1032). No longer tied to the missions upon the expulsion of religious orders in the late 18th century, native people had retreated inland from along the rivers into the more remote regions of the basin. Second, establishing labor in the basin involved high set-up costs; not only did labor have to be recruited from a great distance but new arrivals often faced a tough and alien environment with only the barest of infrastructure and services at their disposal. Finally, maintaining labor also was expensive, as provisions generally had to be transported over hundreds, even thousands of miles under unfavorable conditions. Certain social factors, as discussed below, made labor even more scarce than these three reasons would suggest.

Scarcity of labor is a constant theme of the accounts of the period (e.g., see Akers, 1912:67–70 and 81–82). In 1860, at the outset of the rubber era, the non-native population of the entire Brazilian Amazon was about 278,000, of which over 80 percent was found along the lower Amazon River, in the state of Pará. By the peak of the boom in 1910, the population had reached approximately 1.217 million with 30 percent upstream in the state of Amazonas, centered on Manaus (Santos, 1980: 111). Most indicative of the labor shortage was the persistent efforts to recruit workers from many locales (around the basin and internationally), and the provision of resources by recruiters to move and re-establish families in the rubber fields. In addition, high wage levels for unskilled labor in the port-cities of Belém, Manaus, and Iquitos also would suggest that labor was fundamentally scarce in the region: in 1912, real wages in the Amazon were slightly more than eight times those in Asia, and this is after the price decline of the 1910s was well under way (Akers, 1912: 102).

Labor was drawn from several distinct pools, each varying significantly according to origin and labor experience, motivation for involvement, and their basic material conditions or opportunity cost that conditioned participation. At least five distinct sources of labor can be identified:

1. *Caboclos*: This riverine group was comprised of peasant backwoodsmen of primarily mixed European and native ancestry, though it also included free and escaped slaves. Many caboclos along the Lower Amazon had been involved previously with other extractive products, had gained considerable knowledge and experience in the area, and acquired squatter's rights to the rubber lands they worked.

2. *Cearense*: From the Northeast of Brazil, this group became a particularly important source of labor for the Upper Amazon following the droughts in Ceará of the late 1870s. Some Cearense came with capital, drawn by the opportunity to become patrons or traders; the majority, desperately poor, came with minimal experience or knowledge of jungle life, often under Brazilian government sponsorship, to work as tappers.
3. *Andean Highlanders*: This group of mestizos migrated down from the Sierra and high jungles of the Andes, some as former cinchona bark traders, seeking fortunes in rubber. Many had managerial experience and capital available for investment. Several of the famous rubber barons (e.g., Arana, Suárez, Fitzcarrald) were from this group. Tappers who came under contract (*enganche*) were from poor but generally less desperate straits than the Cearense tappers.
4. *Internationalists*: Most international migration came from among the Andean countries, but fortune seekers also came from Europe and North America to become patrons and tappers. Barbadians were brought to work as overseers and laborers on estates. Moroccans, Lebanese, and Sephardic Jews were well-known for their role in trading, both as itinerant traders along the river and as agents of export houses in the major city ports. Portuguese were perhaps the most evident in the cities and the principal foreign group in the Brazilian Amazon (Burns, 1965).
5. *Amerindians*: Native people from inland regions and far upriver were drawn to the industry by the opportunity to acquire trade goods (e.g., steel axes, machetes) that would improve their subsistence and security position or, in some cases, by force as in the brutal slaving raids (*correrías*) of the Upper Amazon or in the "Devil's paradise" of the Putumayo region (Hardenburg, 1912). Where participation was voluntary and sporadic, native peoples appear to have maintained their tribal groupings, though disease and persecution threatened the viability of tribal life: those groups that were incorporated fully into the industry often became acculturated (Murphy and Steward, 1956).

Although labor was fundamentally scarce during the boom, such labor was, in fact, much more mobile than previous research on era suggests. Prior observers emphasize the tying of labor to estates by "debt-bondage" and the infamous "Rules of the Rubber Fields" which aimed at holding labor on particular estates and patrons. First-hand accounts from the era, however, suggest a rather different picture, one of considerable labor mobility. The scarcity of labor in the basin sustained the prospects of finding employment for inmigrants who provided the bulk of manpower in rubber extraction and often moved several times within the region (see Woodroffe, 1914; Weinstein, 1983a,b; Lange, 1912; Akers, 1912). Generally, estate owners and traders were unable in most regions to enforce the "Rules of the Rubber

Fields" (Weinstein, 1983b:30). Estates experienced considerable turnover of labor, as tappers changed estates, went to work in caucho, the city, or in ancillary activities. Tappers changed estates for a variety of reasons, from seeking the protection of a new patron from an old patron bent on reclaiming a debt to the prospect of working on better estates. Estate owners had difficulty attracting tappers where trees were no longer highly productive (*India Rubber World*, vol.25, no.2, Nov. 1, 1901, p. 46).

Seasonal migration of tappers to the cities and towns during the rainy season and their return when the rivers and rains subsided, often to the same estate and patron, also was very common. Early in the boom, Cearense would leave their homes and go up to rubber fields, gather and sell rubber, and take home pay to families. Over time, many of Cearense who originally migrated annually settled in upriver areas (*India Rubber World*, vol. 30, no. 3, June 1, 1904, p. 298, and vol. 41, no. 2, November 12, 1909, p. 45). In addition, as the boom developed, rubber workers commonly migrated from estate to town according to the work season. Herrera (1914:49) writes,

> En toda la Amazonía las exigencias de la industria gomera ... hace que las poblaciones de las ciudades principales como el Pará, Manaos é Iquitos; y de otras secundarias como Santaren [sic], Obydos, Teffé ... Contámana, etc ... tengan dos períodos bien distintos, que son: uno de completo lleno en su población; y otro de mengua en ella: coincidiendo el primero con ... las lluvias frecuentes, las crecientes de los ríos y las innundaciones de los terrenos bajos ... saliendo entonces todo el personal seringuero, y cauchero, después la *zafra*, del fondo de las selvas a las cuidades ....

The impact of such migration is reflected in reports of the variable size of Amazon river towns and cities, according to whether the tappers were in town. U.S. Consul Fuller (1912:913) writes, "[t]he population of Iquitos is variously estimated at 12,000 to 15,000. In March, April, and May, when the rubber gatherers are not at work in the forest, it will run up to 20,000." The population of Contámana (Ucayali River, Peru) would swell from about 500 during low water to 1500 when the tappers returned (Fuller, 1912:926). Reports of such migration are also available for the Yavari River where Lange (1912) spent the off-season (i.e., five months) with tappers in Remate de Malas (Benjamin Constant) observing their habits and then accompanied them back upriver to their respective estates. Such evidence suggests that seasonal migration from the rubber fields was an important phenomenon—one that has been overlooked in the literature—and supports our contention that labor was relatively mobile during the boom. Undoubtedly some exceptions occurred, as personal liberties were seriously infringed upon, for example, along the Putumayo River, but such cases were limited most likely to the remotest reaches of the basin (see Weinstein, 1983b:30).

CAPITAL

Capital was also fundamentally scarce in Amazonia but less so over the course of the Rubber Boom, with international capital flowing into the region, drawn to the opportunities offered by the wild rubber industry.

Rubber extraction and transport were relatively capital-intensive activities, at least in terms of working (rather than fixed) capital. Tappers and traders needed financing in the form of provisions for several weeks to a full season depending on the remoteness of their operation. Most capital was embodied in transit (i.e., basic goods being delivered and rubber being gathered and brought to port), or in transportation infrastructure (e.g., boats, warehouses, and docks). Although the extractive technology used was rather minimal in value, the setup of the estate (e.g., estate demarcation and estrada layout), the basic provisions for the tapper, and the boats which conveyed provisions and rubber represented substantial investments.

Prior to the boom, capital accumulation in Amazonia was limited to Belém, at the mouth of the Amazon, where significant trade and commercial agriculture were conducted (Weinstein, 1983b). Until the 1850s, the internal waterways of the region were closed to international shipping, most commerce being undertaken by petty traders. Capital was scarce among most tappers as well as many patrons, particularly those interested in expanding upriver, and probably was most readily available downriver near Belém where more abundant capital and a firmer base of property rights and relations made borrowing against the value of rural property more likely.

Much of the capital in the rubber industry, like labor, was mobile and perhaps more highly so. The very mobility of capital allowed the industry to expand into the farthest reaches of the basin, incorporating uncharted areas by sending men with provisions advanced on credit to work rubber. Moreover, capital came from a wide range of agents, from large foreign trading houses to local traders, which helped to keep credit markets competitive. Evidence of this mobility of capital is also present in the transaction among trading houses of debt contracts with traders and patrons, in which trading houses would augment their sourcing of rubber from a particular area by buying out the debt of a series of patrons along that river. Furthermore, trading houses and traders could choose to seek new areas in which to finance extraction if existing areas were proving unprofitable. Again, exceptions probably occurred but only where one or a small groups of traders effectively controlled exchange (e.g., remotest reaches) and if it was clearly in their interest to limit the mobility of finance and trading capital.

Relations of Extraction

The specific forms of relations that developed around land and rubber trees, labor and capital were strongly shaped by the relative abundance and mobility of each factor as well as by the related risk and transaction costs. We begin here by considering the nature of property relations that emerged during the boom, and then examine labor arrangements and capital relations.

Property Relations

Property rights during the Rubber Boom in Amazonia were constituted essentially around the extracted rubber, the land and rubber trees, and watercourses of the basin. In principle, the waterways of the region during the boom remained in the hands of the state and were open to free passage by domestic and international parties. The latex of the rubber tree was considered to be the property of the extractor, and such ownership was distinct from rights to land or rubber trees (Russan, 1902:6). Rights to rubber lands were allocated by the state in the form of concessions and land grants and in most areas squatter rights were recognized at least until more formal forms of tenure were introduced. Land taxes were minimal (see Pearson, 1911:77,142,160).

The land tenure relations that developed during the boom were profoundly influenced by the type of available rubber and the prospective location of extraction. In the case of caucho, much of the rubber was extracted on state lands with no formal claims made by the caucheros or required by the state. Caucho was gathered by highly mobile bands of men who often did not recognize even international boundaries in their pursuit of wild rubber.[12] As a highly transient land use, little *in situ* investment was required and permanent settlement or agriculture were unnecessary. Nevertheless, caucho was also extracted on formal estates as well as from areas subjected to informal claims. Such properties often were very large, in some cases encompassing areas of hundreds of thousands of hectares of hinterland and large groups of native people, and are remembered as those of the rubber barons. The Peruvian Amazon Company (under Julio C. Arana), for example, held definitive title to 5.750 million hectares in the Putumayo region (Schurz *et al.*, 1925:364). Ordinaire (1892) notes that rivers were known in remote Peruvian jungle by the names of the caucho bosses. Where estates encompassed much or all of a remote tributary, the waterways were sometimes considered to be "closed" to trade: exclusive rights to trade and transportation on closed rivers were held by the owner or claimant. Such closure was costly to maintain as the Rubber Boom expanded into these far reaches, and such properties despite their immense areas probably contributed only a modest proportion of the total caucho extracted.

In contrast to caucho, hevea was tapped on estates that lined the rivers and often did not extend more than 10 km inland. Hevea estates were held by a variety of claimants from independent tappers and patrons who contracted other tappers, to commercial firms (often foreign) which hired workers and overseers to run the rubber estate. In many cases, the

---

[12] See Lange's (1912:367–389) vivid account of a deadly struggle between a marauding band of Peruvian caucheros and Indians along the Brazilian side of the Yavari River. Woodroffe (1914) reports gathering rubber in Ecuadorian territory from Peru.

TABLE 3. THE RUBBER ESTATES OF THE TAHUAYO RIVER BASIN, PERU, DURING THE RUBBER BOOM

Estate Name	Map Reference Number	Owner's name	Total Area (hectares)	Number of Estradas		Rubber first extracted	Year of survey
				FINE RUBBER	WEAK RUBBER		
Constancia	**8**	Rafael Pinedo Ríos	3008	0	30	1895	1912
Actividad	**9**	Rafael Pinedo Ríos	4854	54	0	1877	1912
San Antonio	**10**	Juliana Hidalgo Vda. de Ríos	215	2	0	na	ns
Tahuayo	**11**	F. Ruíz	ca. 5000	na	na	<1895	ns
Florida I	**12**	Josefina Flores	ca. 125	na	na	<1897	ns
Florida	**13**	Josefina Flores	2025	0	43	<1897	1916
El Trabajo	**14**	Rafael Pinedo Ríos	8105	18	16	1877	1912
Boa Vista	**15**	José Atúñez Bicerra	20980	70	120	<1887	1913
San Juan de Río Blanco	**16**	Felipe Santiago Flores	988	na	na	na	ns
Monte Cristo	**12**	Adelina Ruíz de Flores	18	na	na	na	1923
Alto Cerrillo	**18**	Angela Vásquez Vda. de Flores	48	0	2	≤1906	1916
Tipishca	**19**	Carlos G. Prado	9118	40	0	≤1898	1915
Pto. Angélica	**20**	Encarnación A. de Méndez	13097	30	0	1875	1911
Termópilas	**21**	Gavino Chávez	9956	0	65	1894	1912
San Teodoro	**22**	Gavino Chávez	353	na	na	<1909	1911

Notes:
1. na = not available; ns = not surveyed.
2. Data complied from original surveyors' reports contained in property title records held in the files of the Ministerio de Agricultura, Iquitos, Peru.
3. All estates were held by Peruvian nationals, except for Boa Vista (Bicerra - Brazilian) and Tipishca (Prado - Ecuadorian).
4. The owner of the Florida estates also worked as a tapper and was illiterate.
5. Where a property was not surveyed, no record exists of an application for title; such properties were likely held by usufructuary right.

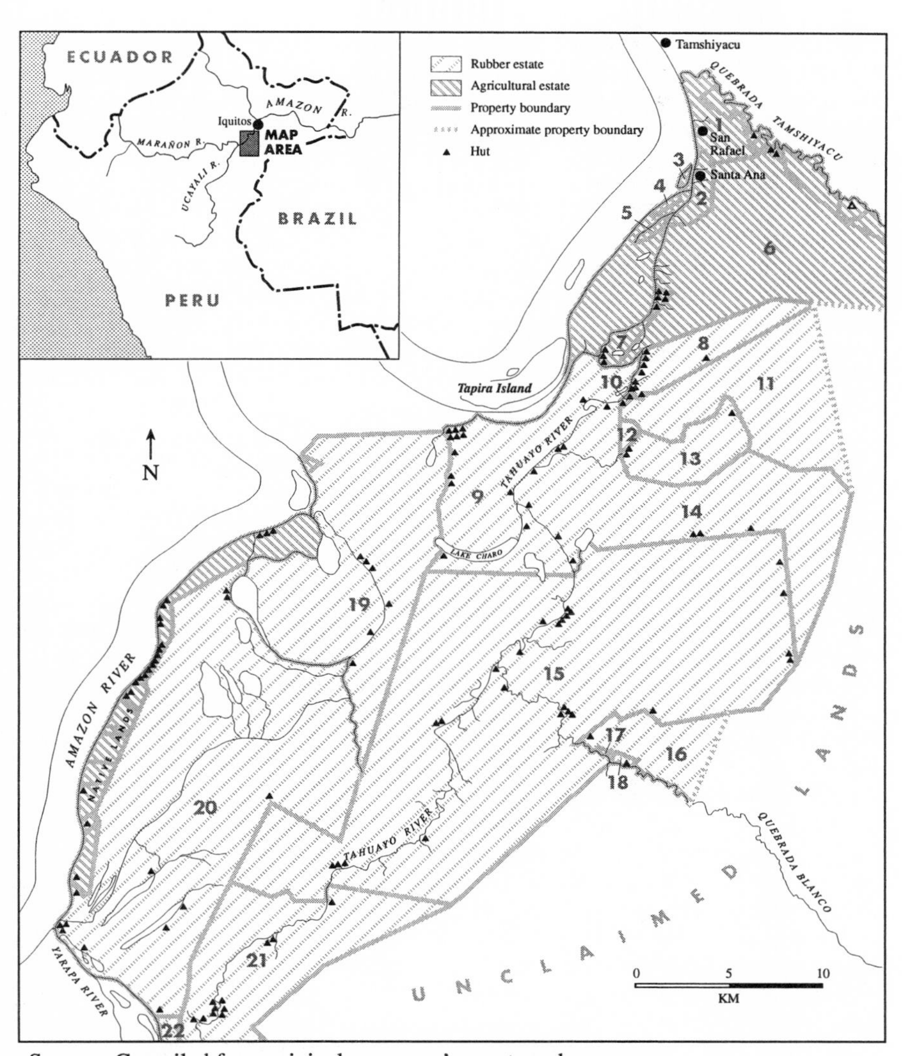

Source: Compiled from original surveyors' reports and maps

Figure 5. Property Holdings at the Peak of the Rubber Boom, ca. 1910, in the Tahuayo River Basin, Northeastern Peru (numbered estates shown on map are described in Table 3)

landscape of property holdings was a mosaic of formal (e.g., concessions and titles) and informal (i.e., possession and squatter rights) tenure, even in areas near the major port-cities. An example of such a mosaic is presented in Figure 5. This map was reconstructed from surveyors' reports to identify individual property holdings along the Tahuayo River, only some 40–80 km south of Iquitos in the Peruvian Amazon, at the peak of the boom. Note the uneven size and distinct purposes of the estates (i.e., rubber versus agriculture), the hinterland areas of unclaimed land, and location of tappers' huts. In general, the focal point of the estate was the owner's hut, usually situated along the river and at the entrance to their estradas. Tappers lived in huts dispersed in the forest and along the river, separated by the extent of their estradas. During the off-season, some tappers would go deep in the upland forest behind the estates or up other rivers to gather caucho, whereas others would go to the nearby town or port-city.

Hevea estates varied considerably in size from the independent tapper with two or three estradas to the patrons and companies that held thousands of estradas. Smaller estates were more common in the vicinity of the major cities of Belém, Manaus, and Iquitos, and along the major rivers; the larger estates were found along the upriver tributaries (Photos 8–11). Weinstein (1983b:45–47) analyzed land holdings near or in the districts of Breves, Melgaço and Anajás (near Belém) and found a large number of small estates (i.e., less than 600 ha, equivalent to not more than a few estradas) apparently worked by independent tappers of humble origins. Estates along the Amazon River in the vicinity of Iquitos contained on average only 21 estradas (derived from Fuentes, 1908:II:81–83). Along the Tahuayo River, rubber estates encompassed between two and 190 estradas, with an average of about 30–50 estradas/estate (see Table 3).

In the more remote areas of the Amazon basin, hevea properties tended be more extensive in area. Schurz *et al.*'s (1925:288) map of rubber properties in Bolivia (Province of Vaca Diez, Department of Beni) after the peak of the boom suggests the grander scale of such estates in this more isolated region, though smaller properties are noted along the main rivers (i.e., Beni and lower Madre de Dios). The Bolivian Amazon, along with the Acre region of Brazil, boasted some of the largest hevea estates in Amazonia. Nicolás Suárez, for example, owned one of the largest Bolivian estates, encompassing some 20,758 estradas (Fifer, 1970:141). The very large estates were the domain not only of certain individual, but also foreign firms. Companies such as the Comptoir Colonial Français, Inca Rubber Company, Inambari Pará Rubber Estates, Ltd., acquired concessions to substantial areas of remote forest for hevea extraction. Most such firms, though, were late-comers and were decidedly unsuccessful in their operations. In fact, the bulk of hevea produced during the boom may well have come from the numerous, small estates of independent tappers and patrons rather than from large and less productive estates; this proposition, however, remains to be tested empirically.

Photo 8. Rubber Estate and Tappers near Iquitos, Peru, ca. 1910
(from Pearson, 1911:153)

Photo 9. Rubber Estate, "Sebastopól," Purus River, ca. 1905
(from Falcao, 1907)

Photo 10. Rubber Estate, "Sibéria," Acre River near confluence with Xapury River, ca. 1905 (from Falcao, 1907)

Photo 11. Rubber Estate, "Andirá," Acre River, ca. 1905

High variation in land tenure relations during the boom would suggest a rather different role and importance of property relations than suggested in some previous literature. Works that portray the industry as one being dominated by large property holders (e.g., Wolf and Wolf, 1936; Collier, 1968; Bakx, 1988) imply that control over access to rubber lands was of crucial importance in understanding the nature and logic of the industry. Rubber barons, so the argument goes, were able to capture exceptional rents from the extraction and trade of rubber by controlling access to the rubber lands and waterways within their domains. Although access to land was important in influencing the returns to extraction (e.g., for the independent tapper versus the patron-dependent tapper), the existence of other tenure forms, and the prevalence of smaller estates, which probably contributed an important proportion of total rubber production from the basin, suggests that access to land and rubber trees was not the primary determinant of the structure of the wild rubber industry. Far more important was the nature of capital and labor relations.

LABOR ARRANGEMENTS

A variety of labor arrangements was known across Amazonia for the gathering of wild rubber, from independent rubber gatherers who worked alone or in teams to those who relied entirely on a patron-client relationship. The specific arrangements that emerged and came to dominate the industry varied regionally and depended upon the following factors: the endowments, motivations and alternate opportunities open to rubber workers, the property relations under which they worked, and the type of rubber collected. The precise specification of labor arrangements in each region across the basin awaits closer empirical study. Weinstein (1983b,1986) offers perhaps the best example to date of the type of effort needed by contrasting the upriver and downriver experiences of hevea rubber tappers, particularly with respect to tenure patterns and their capture of higher returns (see also Pacheco de Oliveira Filho, 1979). Although a complete map of labor arrangements across the basin awaits further research, labor relations were clearly and strongly shaped by the specific characteristics of available labor, property rights and type of rubber that was sought.

Each of the five distinct labor groups that worked wild rubber—caboclos, Cearense, Andean highlanders, internationalists, and Amerindian peoples—came to the industry with different endowments at their disposal. Many seasoned caboclos, for example, were integrated early in the boom as independent or semi-independent tappers in the downriver areas, where proximity to urban centers gave them a rather favorable position as tappers on their own estradas, in areas with denser trade networks. Landless Cearense, meanwhile, were recruited often to work the large estates of patrons in isolated, upriver locales, where thin trading networks, an unfamiliar environment, and the patron-client relationship made their position much less favorable. The motivations of the different labor groups as well as their opportunity costs on participation in the industry also varied greatly.

Internationalists, for example, came to the Amazon seeking fortune; if conditions were too tough, or the prospect of winning a fortune became unlikely, they could remain just long enough to secure a small nest egg to finance the next adventure (e.g., see Yungjohann, 1989). In contrast, native people, who in many cases had been pushed off of their homelands by the expansion of tapping and trading, sought access to new technologies (e.g., steel axes, knives, machetes, guns) to augment their subsistence/extractive potential and perhaps even to secure their ill-defined property rights from other native groups or rubber men. The motivations that brought native people into the trade, given the limited alternatives open to them, bound them much more to the region than the foreign fortune seeker.

Distinct labor arrangements were associated particularly with the type of rubber extracted. On hevea estates, the scarcity of labor and market features of risk and transaction costs provided strong incentives for hevea tappers and patrons to develop durable relations. Patrons had good reasons for seeking a stable relationship with hevea tappers, one that would endure over several seasons or years. For one, hevea trees are most productive when carefully tapped, and a tapper who viewed his stay on the estate as short-term would be inclined to tap without regard for sustained yields.[13] Moreover, a reliable tapper would serve to keep the patron's costs down because labor was costly to recruit, monitor, and risky in terms of product and labor loss. Also, a tapper who perceived the relationship as ephemeral was more likely to be a source of moral hazard problems, given the relative autonomy and risk involved in extraction. In broad terms, moral hazard refers to *ex ante and ex post* attempts to "lie, cheat, steal, mislead, disguise, obfuscate, feign, distort, and confuse."(Williamson, 1985:51).[14] In addition to moral hazard, patrons had to be concerned for their personal security. Cearense tappers were reputed to be highly skilled and predisposed to using a knife (see report in the *India Rubber World*, vol. 41, no. 2, November 1, 1909, p. 45); any estate manager that treated them unfairly put himself at great risk. Patrons thus had strong incentives for investing in stable relationships with their tappers by offering higher returns and occasional forbearance in return for good behavior.

Tappers also had reason to seek a stable relationship with the patron who offered a good opportunity. Capital-poor tappers required access to a steady stream of provisions in order to tap rubber, and provisions were advanced on credit by the patron. Tappers valued the opportunity to work on estates with high yielding trees, yet recognized the costs entailed with moving to another estate. Moving elsewhere would entail finding a new patron, learning to work unfamiliar estradas in a new locale, and the risk of ending up on inferior estradas in a more remote locale. Moreover, the

---

13 In Malaysia today, over-exploitation of hevea trees is known as "slaughter tapping" and is most common on plantations where tappers pay a fixed rent (Chew, 1991:86).

14 More formally, moral hazard refers to "the possibility that insureds will fail to take appropriate loss-mitigating actions in the insurance interval and will not candidly accept accountability" (Williamson, 1985:51).

riskiness of the environment (i.e., the potential for losses associated with illness, accidents, weather and other factors that could lower his rubber output) provided a strong insurance motive for establishing a stable relationship with a patron. In bad times, a "good" tapper would have the insurance of the patron as a source of credit for food, equipment, or medicine that was vital to financial recovery or survival. Such insurance was not available elsewhere in the jungle (again, in part, because of transaction costs), and only a potentially stable relationship that had mutual incentives for building trust would have sustained such an insurance role.

In contrast to hevea extraction, the one-shot nature of caucho collection afforded a much wider range of labor relations, from cooperative groups of caucheros who divided up their rubber earnings (see Fuentes, 1908: I:214) to native people working under oppressive conditions (see Hardenburg, 1912). Between these extremes was a rich array of alternative arrangements: expeditions by teams of patrons and peons, in which the latter received only a small share of the rubber delivered to the patron with the rest being set against their debts; barter exchanges between caucheros (or traders) and native people; and, independent hevea tappers working caucho in the off-season.

The greater potential for coercive relations of labor in caucho than hevea was the product of distinct property relations and the labor incorporated into extraction. Caucheros worked deep in the interfluvial upland forest, often well beyond the reach of government administration and defined property rights. These upland areas had become the refuge of indigenous peoples whose riverine populations had been decimated by disease, reorganized by the missionaries, and persecuted by river traders in earlier times. For the most part, native people were considered by the invading caucheros to be a security threat rather than as potential workers or traders, and Amerindians were ruthlessly persecuted in raids (correrías), particularly in the Upper Amazon. Where trading with native people occurred, great discrepancies in the value of trade goods and caucho are noted. Native peoples along the Putumayo, for example, are reported to have received typically one-eighth of the value of caucho in trade goods (Woodroffe, 1914:160). Such a large spread was very common and reflected in part the fundamental differences in the valuation of caucho and trade goods between the two trading parties. To the Indians, caucho had very limited value other than for trade whereas manufactured goods, which could only be secured through trade, had very high value for their potential use. In contrast, the caucheros valued scarce caucho much more highly than trade goods which were relatively abundant to them. On these grounds, then, trade deals were struck between caucheros and Amerindians.

Some of the most extreme cases of direct abuses of labor in the extraction of caucho are found where the state, in an effort to expand its control over poorly defined boundary areas, granted wholesale administrative authority to private interests. The most notorious of such cases was that of "the river that God forgot"—the Putumayo River—under Julio Arana and the Peruvian Amazon Company (see Hardenburg, 1912; *House of*

Commons Sessional Papers, 1912–13; Collier, 1968; and Taussig, 1984, 1987). In relative terms, however, such extreme cases though appalling in terms of human suffering probably accounted for only a small share of the rubber extracted in Amazonia and most were relatively short-lived affairs. A fuller understanding of labor relations depends not only upon much needed empirical work that maps out relations across the basin but also a closer look at the nature of credit relations with patrons and tappers.

CAPITAL: THE DEBT-MERCHANDISE CONTRACT AND TRADE RELATIONS

Scarce capital was essential to the wild rubber trade and became increasingly available in Amazonia as rubber prices rose with expanding international demand during the boom. Collectors of caucho needed capital to recruit a team, purchase several months of supplies and trade goods, and to secure transportation to the often remote collecting area. To finance a team of 15–50 caucheros in Peru, for example, required 20,000–50,000 soles ($9,750–24,333) in ca. 1906 (Fuentes, 1908: 214). When caucho was gathered on extensive properties, additional investment was required for local infrastructure. In some cases, the investment was massive by contemporary standards. The Peruvian Amazon Company invested £1,000,000 ($4,866,500) in the Putumayo property in establishing villages, outposts, trails, and steamer stops as well as in the steamers that plied the waters of this closed river. On the Ucayali River, C. F. Fitzcarrald is attributed with creating an extensive though relatively ephemeral system of some 200 steamer stops and villages linked by his own steamships.

Hevea tapping required more sunk capital than team-based caucho gathering from unclaimed forest but was a much less risky venture (Plane, 1903:317). The primary costs associated with hevea extraction lay in setting up the estate (e.g., costs of surveying estate boundaries and laying out the estradas), in recruiting labor from the river port-cities, Ceará and elsewhere, and in outfitting the tappers. To open a 50 estrada estate typically required an investment of about £1000 ($4,867 US) (Woodroffe, 1914:215–6). The cost of labor recruitment and transport ranged widely depending of the distance to the rubber fields, from £20–£100/tapper. Each worker would receive rubber extraction tools valued at about $100–250, and supplies for the six month season worth $300–500 (derived from Akers, 1912:69; Santos, 1980:149,166; Loureiro, 1986:109–10; and, Woodroffe and Smith 1915:52). In both hevea and caucho extraction, advanced provisioning was essential to the efficient collection of this dispersed, low yield resource, and provisions constituted a major share of operating costs.

A variety of investors—from international and domestic firms to local participants—provided the capital needed to procure wild rubber. In general, few investors were willing to attempt direct capital investment in rubber extraction, though some did with minimal success,[15] preferring instead to

[15] See DeKalb (1890:192); *India Rubber World*, vol. 24, no. 5, August 1, 1901, p. 328; Russan (1902:6–7); Cooper (1917:327); Melby (1942:464–67); and Weinstein

provide credit to enable tappers to extract rubber and traders to bring the rubber down river. An individual's position within the trade was strongly influenced by the amount of available start-up capital: those from Belém or the high jungles and sierra of Bolivia and Peru who brought some capital with them often became patrons, claiming properties against which operating capital could be borrowed, whereas those with only their labor and no claims to land were more likely to become tappers. As the demand for wild rubber expanded and prices rose, credit flowed readily from international firms to the export houses of Belém, Manaus, and Iquitos. Such credit was used to purchase provisions that were disbursed into the hinterland through the aviamento trade system which linked the urban export houses and rubber receivers to river traders, patrons, and tappers in the forest.

The entire wild rubber trade depended fundamentally upon an agreement between participants by which credit was advanced in return for future delivery of collected rubber. Merchandise, not cash, was the preferred form of credit: cash was scarce and less useful than provisions at any distance from urban markets, more sensitive to inflation, and provided few incentives or guarantees of contract security. By this arrangement, referred to here as the "debt-merchandise contract," the rubber worker was indebted to the patron in the amount of the supplies (as well as initial transport and set-up costs) provided for the season or trip. Typically, provisions were marked up by the patron who also discounted the value of the tapper's rubber. If the debt was not entirely canceled upon delivery of the rubber, the collector's remaining debt would be payable during the next round of gathering and the debt could effectively persist for many seasons, trips, or even years. Payment of the outstanding debt by a third party was sufficient to transfer the tapper (and his debt plus a premium) to another patron. Like the rubber worker, both the patron and the trader received provisions through forward billing from the export houses; again, supplies were marked up and rubber was discounted at each intermediary step, forming a bridge of debt that extended from the port-city to the rubber fields.

Few features of the wild rubber industry have been as strongly criticized by commentators on the era as the debt-merchandise contract and the pricing structure of the aviamento trade system. Both features are commonly viewed as having been singularly exploitative of rubber workers. Observers commonly refer to the contract and trade system as being "inefficient," "precapitalist" (Weinstein, 1983b), "coercive" and a form of "debt peonage" or "slavery" (see Wolf and Wolf, 1936:34; Chirif and Mora, 1980:285; Coates, 1987:94–6; Hecht and Cockburn, 1989:62; Bakx, 1988:147–8; and, Fearnside, 1989:388). Moreover, such features are seen to be the primary impediments to capital accumulation, expansion of rural effective demand, modernization of the wild rubber industry, and broader, more sustained economic development (see Weinstein, 1983b: 263–63; Weinstein, 1983a: 135–36; and Schmink and Wood, 1992:45).

(1986:67–68). For accounts of the difficulties encountered by more recent efforts in Amazonian rubber development, see Dean (1987).

How did such nefarious capital and trade relations arise and come to grip rubber workers as virtual "slaves"? In general, rubber workers are seen to have been drawn into debt relations by their poverty, ignorance, and preference for personal autonomy, and held firmly in the bonds of debt by patrons and traders who monopolized exchange and captured excess profits from their labor. The primary evidence cited of exploitation, surplus drainage, and trade monopolization is the persistence and pervasiveness of the debt-merchandise contract and the pyramid structure of pricing for provisions and rubber down the chain from export houses to traders, patrons, and tappers. The tapper, at the bottom of the price pyramid, bore the worst terms of trade, paying in some cases 250–500 percent over city prices for basic supplies and receiving less than 50 percent of the prevailing city price for his rubber (see Figure 6). Such pricing, observers argue, kept tappers' earnings barely above subsistence and allowed for surplus drainage to rubber industry elites and eventually out of Amazonia entirely, via profit remittances and import expenditures.

The prevailing view of the debt contract and trade relations, which relies heavily upon the assertion of widespread trade monopoly, can be challenged on two grounds. First, as we argued in Chapter Three, considerable evidence exists to suggest that competitive rather than monopolistic trading was the rule during the boom. Few barriers existed to keep prospective traders from entering the trade and opportunities for trading in the vast basin were abundant. Numerous accounts are available from the period that note the success of itinerant traders in exchanging their wares for hevea and caucho, and of the frustration of foreign and local concerns in limiting the leakage of rubber out of the traditional aviamento system (e.g., see DeKalb, 1890:192). Itinerant traders flourished around the port-cities, towns, and steamer stops. Vociferously denounced as pirates and unscrupulous inter-lopers, efforts to curb itinerant traders were generally unsuccessful in part because of support given by urban-based merchants (Weinstein, 1983b: 51–2).

Of course, some of the more remote rivers were indeed effectively closed to competitive trade. However, in general, many of the most productive rubber rivers were open waters, and competition among traders effectively undermined efforts to monopolize trade on open rivers. In Peru, infamous for its distant rivers controlled by rubber barons, at least 78 percent of the rubber exported near the peak of the boom came from open, not closed rivers (see Fuller, 1912:924–37). Trade monopolies were not only unpopular in the port-cities where commerce centered on the provisions trade but they were expensive to maintain given the high cost of enforcement, the low productivity of labor, and the high mobility and persistent threat posed by itinerant river traders. Effective competition, stimulated by itinerant traders, rather than brute force was probably the more successful strategy for maintaining flows of rubber and merchandise within credit lines.

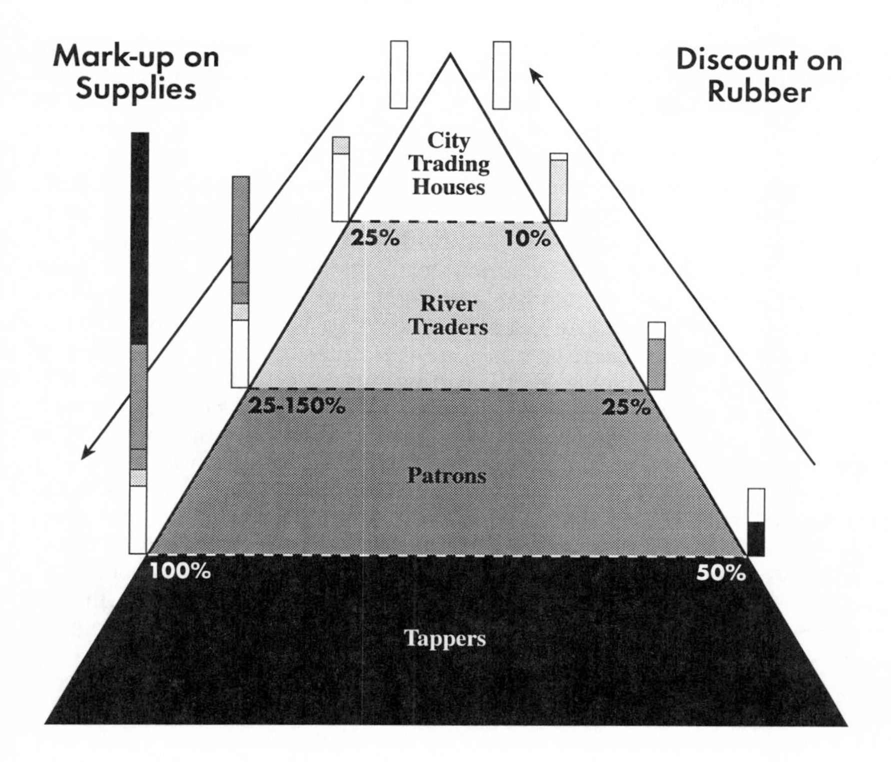

Figure 6. Example of Pyramid Pricing Structure of Debt-Merchandise Contracts During the Amazon Rubber Boom

Second, both the debt-merchandise contract and pyramid pricing structure were prevalent throughout Amazonia, whether rivers were open or closed to outside trade. Whereas monopolized trade on a closed river may indeed have enabled traders to demand higher prices on goods, offer lower prices on collected rubber, and thus to hold tappers in debt relations, competition along open rivers would have effectively improved prices for patrons and tappers and eroded any individual trader's tight monopoly position. If monopoly over trade was necessary to maintain debt relations, as observers suggest, then such competition would also undermine the viability of the debt-merchandise contract; yet such contracts were common along open rivers. Moreover, if trade power was crucial in determining the structure of pricing, price margins and monopoly returns should have been lower, not higher for local participants (i.e., patron-tapper and trader-patron) than purportedly more powerful concerns (e.g., rubber receiver-trader, or export house-receiver). Clearly, trade monopoly is not sufficient to explain the persistence of the debt relation and the geographical pervasiveness of the pyramid pricing structure of the rubber trade.

A more compelling interpretation of the debt relation and pricing structure lies in the specific characteristics of the environment in which wild rubber was extracted and the respective needs of industry participants. In our view, the debt-merchandise contract emerged as the dominant arrangement for rubber extraction because no other alternate contract could meet the needs of creditors and borrowers at lower cost in this high risk, high transaction cost environment where both labor and capital were scarce. In choosing the terms of an arrangement with a hevea tapper, the patron had several objectives: he sought to secure a steady stream of rubber; to avoid the significant expense and risk entailed in engaging replacement tappers; and, to ensure that the rubber trees on the estate were tapped sustainably. To the tapper, the patron represented both a source of credit in a capital scarce environment, without which rubber could not be tapped and fortunes made, and of much needed insurance. It was, after all, the capital-poor tapper who bore greatest risk to his person in tapping wild rubber from the forest. The advance of credit and goods represented not only an obligation to be met by the tapper (i.e., a debt) but also a commitment on the part of the patron to the tapper to enable him to carry over his obligation (i.e., insurance). In return, the patron was relatively assured of a reliable stream of rubber from a tapper whose incentives were to tap conscientiously so as to ensure access to additional credit. By joining in a mutual commitment of credit for rubber, patron and tapper met their respective needs and ensured that wild rubber would be procured at low cost per unit capital and per tree.

In caucho extraction, the debt-merchandise contract often included a share of the final product for the gatherer. Unlike hevea, caucho could not be sustainably tapped and the tree was felled to extract the latex. Credit was advanced by the patron in the form of supplies to team members and the harvested caucho was sold or exchanged for goods along the rivers or in the port-cities. Whereas a team of international fortune seekers may have kept a high share of the product, the indebted local worker probably retained a smaller portion of the surplus. For tribal peoples engaged in the trade, advanced provisions were not essential, as they had access to a secure local source of subsistence foods and materials, and trade was based on barter.[16] The "credit plus share" contract enabled the caucho worker to travel and subsist for long periods in the forest and provided incentives to maximize the harvest and ensure delivery of all of his product.

The relative efficiency of the debt-merchandise contract can be appreciated by considering alternate contractual arrangements by which hevea rubber could have been harvested. Among such arrangements, most of which were probably attempted but with only limited success during the

---

[16] In some cases, tribal people were bound to the patron or trader by extra-economic forces. On the Ucayali River (ca. 1880-1898), natives worked caucho under C.F. Fitzcarrald who they regarded as "El hijo del Sol" (Reyna, 1942: 22). On the Putumayo, mutual fear and extreme violence linked native collectors and overseers, see Taussig (1984). Nevertheless, in both cases economic relations persisted as tribal natives exchanged caucho for goods.

boom, are wage tapping (fixed wage paid to tapper), rental (tapper rents estradas), commission tapping (tapper paid per unit of rubber tapped), and share tapping (percentage of rubber collected to tapper) (see Chew, 1991). Wage and share arrangements would (and did) suffer serious problems with monitoring the output and ensuring delivery of all of the rubber tapped, because tappers would be sorely tempted not to report all their extracted product, selling instead to itinerant traders, other patrons or even other tappers. Sufficient objective risk pervaded the environment so that tappers could readily justify low yields whatever their true cause. Under an affordable short-term rental contract, the patron would need to monitor output, not primarily to ensure his return but rather to protect his trees from overtapping and destruction. Under a commission contract with no credit provision, most tappers would still have needed supplies before any trees could be tapped. As such, commission arrangements would evolve into debt-merchandise relations, given the need for advanced provisioning and the advantages of provisions over cash. In all cases, additional costs would be incurred by the imperfections embodied in these alternate arrangements, making them less competitive with respect to the debt-merchandise contract. No other arrangement could have met the needs of patron and tapper and provided rubber less expensively per unit of capital invested and tree tapped. For this reason, the debt-merchandise contract was the most pervasive and persistent arrangement used in the extraction of wild rubber during the boom.

The historical record following the boom is revealing here on the importance of the price-cost environment and relative factor scarcity in shaping capital-labor relations. When rubber prices plummeted in the 1910s and 1920s, other arrangements than the debt-merchandise contract became more feasible for the extraction of wild rubber. Reports on conditions in the rubber fields during the 1920s indicate that the debt-merchandise contract in many areas was replaced by share and fixed rent contracts (see Schurz *et al.*, 1925:27). With the dramatic fall in rubber prices, many tappers turned of necessity to subsistence production of agricultural food crops thereby reducing the need for advance provisioning. Other forest products and commercial agricultural crops, such as Brazil nuts (Brazil), and cotton (Peru) became competitive and estates diversified production (for a case study from Peru, see Coomes, 1995). Diversification and the rise of agriculture effectively redefined the relative importance of the factors of production: land became the primary determinant of returns and control over access to land, rather than capital or labor, now shaped labor relations. On many estates, debt-merchandise contract tappers became tenants or share tappers and this transition was a direct response to changes in the prevailing economic environment.

The pyramid pricing structure of the rubber trade, like debt-merchandise relations, also can be understood in an alternate manner. Our explanation does not rely on the assertion of widespread trade monopoly but instead on the crucial role played by risk in Amazon River trading. Price margins, in our risk-based explanation, can be conceived of as interest rates on merchandise-based loans. Smaller price margins reflect lower interest rates on loans from, say, the export house to the traders, with larger margins reflecting the higher rates on loans between patrons and tappers. Creditors (e.g., patrons or traders) can be seen as building in a risk component to loans to ensure that the expected return would equal some alternate, safe investment. Under conditions experienced in rubber extraction during the boom, substantial risk premiums had to be build into their price margins.

The impact of risk on mark-ups on advanced provisions can be best illustrated with a hypothetical example. Suppose that a patron anticipated that two of ten tappers would die, desert, or become incapacitated by illness. Such a rate would have been considered as low for tappers during the boom (see Chapter Three from p. 44). What interest rate must the patron charge to secure a return of 10 percent on the investment? The nominal rate of interest (i) would be calculated as,

$$i = (d+u/1-u)$$

where: d is the desired rate of return on investment
u is the expected rate of loan default

Administrative costs associated with the loan (i.e., opportunity cost on lender's time) are assumed here to be zero. To secure a return of 10 percent, the patron would thus need to charge a rate of 37.5 percent to each of the ten tappers. An administrative cost of 20 percent would raise the rate to 62.5 percent.[17] If a third tapper was then to default under such an arrangement, the patron would lose not only the principal but the opportunity cost of foregoing returns on a safer investment of the principal. Thus, if the patron was risk averse, the potential for downside losses would make the investment unattractive unless this risk premium was built into mark-up. Such a premium could easily raise the interest rate tappers face to 45 percent (assuming the loss of two tappers) so as to provide, on average, a 15 percent rate of return for the risky investment. What appears to be exorbitant price margin in this example becomes the basis for ensuring normal rates of return in a highly risky environment.

Such is the explanation for high mark-ups in general, but why would price margins increase substantially as merchandise moves down the chain from the export house to the tapper in the jungle? Two other facets of risk are crucial to explaining the pyramidal structure of the trade. First, risk tended to increase with distance into the jungle, and second each participant further down the chain was spreading his risk over a progressively smaller pool of clients. Greater distances meant longer and less frequent trips; hence, larger shipments per patron and tapper were required to ensure the delivery of sufficient supplies. In both directions, the trader faced higher

---

[17] For an explanation of rate determination that incorporates both the risk premium associated with default and administrative costs of rural lending and collecting, see Bottomley (1975:282).

risks, covering greater distances and carrying more cargo per patron (or unit for spreading risk). Obviously, nautical distance would be a rough proxy for the approximation of risk, given that a short river trip through rapids was riskier than a longer, less treacherous trip.

Smaller risk-spreading opportunities also help to explain the particularly high price markups between patron and tappers (see Figure 6). On the estate, the patron was spreading risk over a relatively small number of tappers (e.g., ten) who bore the brunt of hazardous jungle work. He also had little option for diversification both because of the cost of re-transporting the commodities and the transaction costs of having another party, at a distance, oversee their use in a similar, risky activity. If each trader, however, worked with ten patrons along a river, then each trader had his risks effectively spread across 100 or more tappers. If the export houses, in turn, worked with ten traders, then their risks were spread across 1000 tappers. The greater the pool for spreading risk, where outcomes are essentially independent, the lower the risk of substantial losses.[18] Because the greatest credit risk along the chain was carried by patrons, it is understandable that patrons charged a higher interest rate to the tapper than the patron paid the trader. In this way, price margins increased out from the port-cities into the hinterland and down the chain from trader to patron and tapper.

Our alternate interpretation of the debt-merchandise contract and the pyramid pricing structure stands in sharp contrast to prevailing views of these two crucial features of the industry. Not only does this interpretation offer a more robust explanation of the origins, persistence and pervasiveness of these two crucial features of the wild rubber industry, but it also provides a stronger foundation upon which to assess other issues such as rubber worker exploitation and failed economic development in the region (see Chapter Five). More relevant to our purposes here, however, this revised view is helpful in considering the level and distribution of financial returns to industry participants.

---

[18] This notion underlies the design of all insurance systems. When a large number of independent alternatives are pooled, the risk of large losses is reduced to a small probability event, because the joint probability of independent events is the product of the individual probabilities. In our example, if each tapper has a one in five chance of dying or deserting, and if these are viewed as independent phenomena, then the risk to the patron of losing his entire working capital investment is 0.2 to the tenth power or one in ten million. For the trader with 100 tappers in his scheme, the probability of total default will be minuscule. The greater the number of independent events, the more the range of outcomes will be squeezed around the mean, by the reduction in probability of big losses or big gains. This will, in turn, lower the variation of expected returns and reduce the risk premia necessary to attract investment or participation.

Financial Returns

Much evidence exists in the secondary literature to suggest that the rubber trade was a rent-filled industry with considerable potential for the accumulation of capital in the region. Santos (1980:299) has estimated that the Gross Domestic Product/capita in the Brazilian Amazon rose dramatically during the boom, from $49/person (1840) to $323/person at the peak of the boom in 1910, and then dropped with the collapse of rubber prices to $74/person in 1920. Sizeable fortunes are reported to have been amassed, often by local men of rather humble beginnings (e.g., Araújo, Suárez, Arana, Fitzcarrald, etc.) who led the flamboyant lifestyles by which the Rubber Boom is known. The state captured substantial surplus at low cost from the industry through *ad valorem* export and import duties as high as 23 percent and 100 percent, respectively (Pearson, 1911). The persistently high level of rubber output from the region until the 1920s, despite the precipitous and continual decline in rubber prices after 1910 also suggests that the industry was not operating at the margin of profitability in the years prior to the peak, and that very substantial rents must have accrued during the boom.

The problem of how rents from the wild rubber industry were distributed among the participants—from tappers to patrons to traders—is perhaps the most conceptually important yet empirically problematic of the questions to be answered about the boom. This section offers the argument that, on average, substantial surplus was captured locally by patrons, traders, and even tappers in the industry. A paucity of good empirical information on returns to the different participants forces us to rely heavily here upon conceptual argument where hard data would be particularly useful. Nevertheless, the evidence is sufficient to make credible the assertion of local capital accumulation. Further evidence emerges in Part II (Chapters Seven & Eight) where we discuss the patterns of private and public investment associated with the Rubber Boom.

The promise of wild rubber—the "black gold" of Amazonia—was unprecedented prosperity to those willing to invest in the sector. The rubber trade drew men out of local agriculture and other traditional activities. As rubber boomed, the better-off as well as the destitute of northeastern Brazil were drawn to Amazonia as were men from the high jungle towns of the eastern Andes and international fortune seekers. Rubber extraction even drew labor out of activities that were essential to the trade. The U.S. Consul to Belém wrote earlier on in the boom, "... the inhabitants along the Amazon being so absorbed in the production of rubber, owing to its recent rise in value, that labor enough cannot be had to furnish the steamers with wood." (*U.S. Consular Reports*, vol. 12, no. 39, March, 1884, p. 204). As the boom progressed, the movement of men upriver was massive. In one instance, U.S. Consul Kenneday reported from Belém during September, 1903, that 5000 men had left over the previous seven days for rubber fields upriver (*U.S. Monthly Consular Reports*, vol. 74, no. 280, January, 1904, p. 205).

The port-cities of Belém, Manaus, and Iquitos became centers not only for recruiting men for the rubber fields but for the spending of money earned once engaged in the trade. During the off-season, the port-cities and secondary towns swelled in numbers as rubber workers returned from the jungle to seek family, diversion, and investment opportunities. A plethora of the finest imported goods from the world over were available to tempt not only the patron but also the tapper who had come to town with savings accumulated over one or more seasons.[19] As the floodwaters receded, savings shrunk, and debts mounted, rubber workers boarded the river steamers and returned to the jungle estates to join their compatriots who had stayed on during the off-season, working caucho, cutting firewood for the steamers, practicing subsistence agriculture on the upland and high levees, or simply recuperating from the last season.

In setting out for the jungle to work rubber, veterans and new recruits alike left behind in the port-cities and towns opportunities for employment at high (relative and absolute) wages. Rising rubber prices and scarce labor pushed wages up throughout Amazonia: the daily wage in Belém (ca. 1901) was roughly four to five times that in the Northeast (Santos, 1980:113); in Peru, the minimum wage in the Amazon was about double that in the Sierra (*U.S. Daily Consular and Trade Reports*, no. 271, Nov. 16, 1912, p. 856). High urban wages were necessary not only to attract trained labor into the region (e.g., professionals, administrators, etc.), but also to hold untrained labor out of rubber extraction to work the ports and service industries of the cities. An unskilled day laborer in Iquitos during 1910 earned typically $20–40/month, roughly the minimum wage in the late 1980s in the Peruvian Amazon. Even at high wages, labor flowed liberally into rubber extraction—and skilled labor was also drawn out of the urban service sector. At the peak of the boom, tramcar operators in Belém abandoned their posts *en masse* in a rush for the rubber fields (Wolf and Wolf, 1936:66). Clearly for many people, the expected returns in rubber extraction exceeded those obtainable from urban employment.

The estimation of typical returns expected by a rubber worker is highly problematic. Ideally, such an estimate would be based on a full and complete financial accounting of the credit contract over time, from the initial financing to eventual resolution. In our research, however, we encountered no complete tapper-patron financial accounts in the literature (e.g., cost of merchandise, merchandise delivered, rubber credited at relative market prices, possible off-season income or cash sales to itinerant traders, credit advanced for off-season subsistence on the estate or in town) that would allow accurate estimation of tapper returns. Instead, several reports of "average" earnings by tappers were found (e.g., Loureiro, 1986:54–55, 109–110; Woodroffe, 1914:221; Jumelle, 1903:96–100; Lange, 1911; Santos, 1980:166; Arnous de Rivière, 1900:437) in which "typical"

---

19 See reports of tappers spending in a month that which took several seasons to accumulate (e.g., Lange, 1911). For a description of Manaus at the peak of the boom, see Burns (1965).

conditions were not carefully specified. Few reports, for example, are explicit about prices paid and received between trader, patron, and tapper: was the $300 of goods advanced to the tapper the cost to the patron or the charge levied by the patron to the tapper (i.e. including the 100 percent markup commonly charged)? Strong spatial and temporal variations in tree yields, goods and rubber prices, off-season activity, cash sales to itinerant traders, and tapper mortality also limit the representativeness of these estimates. With gross tapper earnings of $350–600/season, assumptions about specific prices, yields, accounting procedures, etc. make the difference between the tapper realizing a healthy economic surplus, breaking-even, or spiraling deeper into debt. A rather unique set of primary data is needed to construct reliable estimates of tappers' returns, preferably from the records of a number of estates or trader zones that would include a clear definition of property relations, prevailing prices, yields, and extractive and productive practices.

In the absence of such information, we build the argument for surplus accumulation by referring to alternate income and savings opportunities available to industry participants. Clearly, the principal motive for prospective tappers to set out for the rubber fields was economic: the minimum expected return to tapping must have exceeded the potential savings of urban or rural laborer. Risks to health and welfare in the jungle were considerably higher than in the cities, and the cost of living rose rapidly with distance from the port-cities. An unskilled laborer would have considered his potential returns in deciding to remain employed or to leave to collect rubber, an activity open to the untrained and capital-poor. In the Brazilian Amazon, rural wage laborers would receive the equivalent of between $24 (Lower Amazon) to $40/month (Upper Amazon) with rations (Akers, 1912:81). Such wages were competitive in the urban labor market and an economizing worker could save perhaps $150–250/year. Returns to wage labor probably were comparable to those expected from rubber gathering, though actual returns to tappers would vary considerably, depending on the tapper's skill and technique, the health of the tapper, rubber prices, and the costs of subsistence as well as natural variations in tree yields, the weather, and the duration of the low water period.

Perhaps the least likely participants to accumulate capital were the destitute Cearense who were recruited and sent at high cost to the remote, upriver estates where they worked under a patron. The cost of living for the tapper in these regions was exceptionally high—often four to eight times that in the port-cities—so that a tapper's supplies for the season may have been worth only slightly less than the value of the tapper's production. Typically, the tapper would remain in debt for at least the first few seasons, paying down his substantial initial loan, and then, depending upon his health, opportunities to work in the off-season, and other circumstances he would either show positive returns or continue in debt. Although the majority probably continued in arrears to the patron, reports do exist of upriver tappers earning several hundred dollars in a good year.[20] Perhaps more important than earning levels was the fact that the persistence of debt did not negate the possibility of capital accumulation. On the contrary, many poor tappers may have preferred to remain in debt so that credit, security, and access to rubber trees were assured while they accumulated assets directly (e.g., receiving a sewing machine, rifle, etc.) and/or indirectly through sales of rubber to itinerant traders.

Some Cearense tappers did cancel their debts and amassed sufficient capital to buy out the patron, go further upriver and stake-out an estate, become a trader, or return to Ceará with the rains. Weinstein (1983b:24) describes one such case of an industrious and fortunate Cearense working rubber along a tributary of the Xingu River. Such individuals provided the inspiration for others who would follow, taking extreme risks with at least partial knowledge (by the evidence of new recruits jumping ship along the way) to work rubber upriver. More than bare subsistence was available in less expensive places (e.g., in or around the cities and towns) and in safer activities (e.g., wage labor or subsistence agriculture). It was the stories of the fortune of few that attracted the poor and destitute to work rubber in the remotest of reaches.

The potential for capital accumulation among independent (or better off) tappers were higher than among the poorest, contract-bound tappers. Those tappers with access to land, either formally or by squatting, could expect higher and more consistent returns than those working on a patron's estate upriver. Independent tappers working small estates (i.e., a few estradas each) around the major port-cities such as Belém (especially in the Islands district), Manaus, and Iquitos, along the Amazon River, and the lower tributaries also benefitted from the lower cost of living. The more efficient producers were able to accumulate some capital even though yields may have declined on these estates where some trees had been worked for decades. Such tappers would have had access to cheaper credit, given their holding of land and proximity to the port-cities, and some tappers were able to borrow directly from the trading and export houses to invest in their holdings or to finance other tappers elsewhere. Again, a wide range of actual returns are expected among independent tappers working their small downriver estates, however, net earnings of up to several hundred dollars per season would seem possible.

Individuals with initial capital were more likely to enter the industry as patrons or traders and receive higher and more consistent returns than capital-poor rubber workers. Although no definitive estimates of rates of capital accumulation among patrons are available, a review of several case studies (i.e., Loureiro, 1986; Santos, 1980; Jumelle, 1903) suggest that a patron in a good year could earn $150–250/tapper/year. Nevertheless,

---

[20] Lange (1912) on the Yavari River in 1910 reported gross earnings of $8–10/day/tapper. Schidrowitz (1911:ca.26) suggests that a tapper could make £50–60 ($243–292) per season in net revenue. A similar figure is reported by M. Bonnechaux ($290–347/100 day season) in *U.S. Consular Reports*, vol. 67, no. 255, December, 1901, p. 575.

accounts are also common of patrons sinking deep into debt, mortgaging and then losing their estates to traders and commercial houses, particularly late in the boom (for example, see *India Rubber World*, vol. 40, no. 1, July 1, 1909, p. 348). Returns to commercial house traders, the most common type of trader, are expected to be somewhat lower than the average patron—perhaps even approaching the equivalent of an administrative wage—but more consistent than those realized by patrons as risk was spread over a larger pool of clients. Despite apparently exorbitant markups, the competitive and highly mobile nature of river trading would suggest that true profit margins for river traders were modest.

Chapter 5

The Performance of the Wild Rubber Industry Revisited

Introduction

Our revised understanding of the organizational structure and logic of the wild rubber industry developed in two preceeding chapters can now be brought to bear on the key problems of the wild rubber industry. Such problems are seen by scholars as being instrumental in limiting the longer term development potential of not only the rubber sector but the entire Amazonian region. Specifically, we revisit three principal failures of the industry—the problems of supply inelasticity, inefficient social relations, and thwarted plantation development. Prevailing explanations of how and why such problems arose are evaluated against our revised understanding of the rubber industry; in each case an alternate, more robust explanation is offered. We then marshall our evidence and arguments to show what previous scholars assert did not happen with the boom—that local participants did manage to accumulate substantial capital through the wild rubber industry—and in doing so, cast serious doubt upon what for many scholars has been the prime explanation for regional underdevelopment associated with the boom. We conclude the chapter by arguing that the focus of debate on performance issues, however engaging and informative, is actually misplaced if the crucial question to be answered is why half a century of Rubber Boom did not lead to sustained economic development in the Amazon.

The Inelastic Supply of Rubber

For early industry analysts, the very real problem of the inelasticity of wild rubber supply lay primarily in the backward and inefficient manner in which wild rubber extraction and trade was conducted. The remedy was seen to lie in "rationalization" and "transformation" of traditional rubber collection and exchange, as was commonly being done in other industries of the time. Recommendations and advice for reform were abundant: workers needed improved technology for harvesting rubber; archaic debt relations had to be broken down and replaced with the discipline of wage work; and, the price gouging middlemen must be squeezed out of the trade. An industry restructured as such would ensure that the supply of wild rubber could shift more smoothly and responsively to changing levels of demand in the key rubber exchanges of Europe and the United States.

Our explanation for supply inelasticity suggests more fundamental causes than the "traditional" or "artesanal" form of organization, causes that reflect the basic geographic features of the region and the inherently product-specific nature of investments in wild rubber extraction.

Market price changes affected the crude rubber supply by signalling to producers in Amazonia whether manufacturers needed more or less rubber. With a price increase, estate owners could attempt to increase extraction from existing properties by pushing further into the forest or by bringing new upriver areas into production. Because untapped and unclaimed rubber trees were abundant upriver, the additional rubber output usually came from the expansion of the rubber trade up into the tributaries and more remote regions of the basin. This upriver expansion typically entailed higher production costs—to recruit, transport, provision, and monitor tappers. Large price increases therefore were needed to drive the spatial expansion of the trade and the increase in rubber supply.

Alternatively, when prices fell, rubber trade participants' capital and labor were tied up in debts, provisions, estates, boats, and warehouses, and could not be rapidly reallocated to other uses. More important, no other activity promised nearly as large an economic return as rubber. Thus, resources were not readily transferred in and out of rubber extraction, and the supply of wild rubber did not contract with small fluctuations in market prices. Even with a huge drop in rubber prices in the 1910s, Amazonian rubber output was maintained at (or above) boom levels for almost a decade. As late as 1917, when prices were already 50 percent lower than their 1910 peak, official rubber exports from Brazil were almost equal to output in 1910 (Santos, 1980:236) (see also Figure 2).

The problem of supply inelasticity, in our view, therefore stems from the dispersed and ever more remote location of wild rubber stands and the relative fixity or specificity of investment in rubber extraction operations, rather than the traditional technology, the relations of extraction, or the trade system. To create a more responsive supply of wild rubber, given the relative efficiency of the debt-merchandise contract and the high returns already being generated, would have required a wholesale transformation of the underlying productive environment.

Inefficient and Resistant Social Relations

Recent literature gives particular emphasis to the views of early writers that a key impediment to the rationalization of the Amazon rubber industry was the persistence of inefficient social relations of extraction and trade. As noted earlier in Chapter Two, the classic view of tapper coercion and patron resistance to change has been challenged by other scholars, though from rather different evidence and perspectives than is offered here. We emphasize instead that labor's fundamental scarcity, mobility, and the high costs of monitoring tapper effort were essential features of production. Such features obviate the classical portrayal of tappers as debt peons, serfs or slaves kept in these relations by coercion.

The more recent view of social relations of extraction (see Weinstein, 1983b:23; Weinstein 1986:70–72), depicts the tapper-trader bond as one founded on the interlocking and distinct interests of each party: autonomy for the tapper, and exchange monopoly for the trader. Two dubious propositions underlie this perspective. The first proposition, that tappers' preference for personal autonomy was paramount, is only sustained by assertion or tautology, and is doubtful given the reportedly wide range of tappers' aspirations (from securing subsistence to making their fortune). Moreover, whereas undermining wage-labor contracts by selling rubber on the sly to "pirate traders" could be interpreted as autonomy-seeking behavior, the same action could be explained simply as tappers taking advantage of a form of labor relations in which their activities were prohibitively expensive to monitor. Whether tapper behavior reflected an overwhelming preference for autonomy or not, pirate sales and other acts of "resistance" could not have been deterred in a wage scheme.

The second proposition is that traders effectively monopolized exchange throughout the basin. This proposition also is questionable in light of our analysis. With many open rivers and numerous itinerant or debt-merchandise traders proffering their wares, with few significant barriers to entry and high costs of attempting to control river trade, the rubber trade would appear to have been fundamentally competitive. Indeed, the trade's pyramid price structure, usually cited as the primary evidence of market dominance and the means by which traders captured industry surplus, is better understood as the product of doing business in a highly risky environment where capital was scarce and labor was essentially unmonitorable. Importantly, such a view does not require rejecting the evidence of "open" rivers in order to assert monopoly power for traders.

A more compelling explanation for the durability of social relations than either the "resistance" or "control" theses lies in the relatively efficient nature of the debt-merchandise contract that bound the patron, trader, and tapper. Other forms of social relations introduced to "rationalize" the industry, such as wage labor, were thwarted not by a *de facto* intra- and inter-class alliances but by their own inefficiency, given the underlying geographic and microeconomic features of the Amazonian wild rubber industry.

When rubber prices collapsed in the 1910s, the economic environment of the region changed fundamentally, and by the early 1920s both trade and labor arrangements had been transformed. In many areas, the trading system of the Rubber Boom era contracted; intermediaries dropped out of the trade chain that linked tappers to export houses. Estate owners consolidated land claims and offered alternative for farming and tapping (e.g., share cropping and share tapping). Owners also encouraged the diversification of production to include a mix of subsistence and commercial agriculture (Schurz *et al.*, 1925: 6–7,26–28; James, 1930:75). The conditions that had favored the debt-merchandise contract during the boom no longer prevailed, and new arrangements resulted between patrons and their workers.

Plantations, Modernization, and the Asian Challenge

The inability to successfully develop rubber plantations in Amazonia during the boom often is seen as a principal reason for unsustained economic growth and stifled development in the region. Our analysis finds that the three contending explanations for unsuccessful plantation development—the *dependista* view of drained surplus, Warren Dean's ecological focus on endemic leaf blight, and the Marxist view of social relations—each contain fundamental weaknesses.

The dependency approach can be faulted simply by pointing to the persuasive evidence that local participants did realize capital accumulation. As will be discussed further below, many groups, including the state, patrons, traders, and tappers, captured substantial rents from wild rubber extraction. Sufficient capital became available to local investors that some attempted to establish rubber plantations in several regions of the basin during the boom. Warren Dean (1987: 45) finds evidence for formal rubber tree planting in the Brazilian Amazon as early as 1865. The first results of plantations area are reported to have been sold as a Manaus consignment of "plantation Pará sheet" in London on December 20, 1907 (see *India Rubber World*, vol. 37, no. 6, March 1, 1908, p. 185). Evidence also is available from the Peruvian Amazon of estate owners attempting, late in the boom, to introduce rubber plantations (see San Román, 1975:133; and Coomes, 1995:111). Such initiatives suggest that local capital indeed was available for investment to transform the industry. Moreover, even if local surplus retention had been limited as the dependistas suggest, abundant foreign investment capital was available to offset that constraint; it was (and has been ever since) precisely the foreign ventures that failed in the attempted shift to plantation production.

The ecologic-economic explanation proposed by Dean is more persuasive for the post-boom period than for the boom itself. The threat of leaf blight was not the principal disincentive to plantation development during the boom. Although rubber planting was more extensive in this period than commonly assumed, and plantations did experience a variety of plant pathologies, our research found no evidence to suggest that potential investors of the time, either international or domestic, knew South American Leaf Blight as a serious danger to rubber cultivation in Amazonia. On the contrary, the fungus which caused South American Leaf Blight was identified by scientists only in the early 1900s and widespread appreciation of its potential significance came even later (Dean, 1987:53–57). The magnitude of the blight was recognized in the ravages and rapid abandonment of rubber plantations not in Amazonia prior to 1910, but in British and Dutch Guiana in the late 1910s and early 1920s (Rands, 1924). As late as the mid-1920s, some U.S. scientists fresh from field studies in Amazonia still could only speculate on the potential threat of the disease to plantation development in the region (see Schurz *et al.,* 1925:100,116; LaRue, 1926:57). That potential was realized much later, with a vengeance, at Fordlandia and other plantation estates (Rands and Polhamus, 1955).

Some estate owners who attempted to cultivate monocrop stands of rubber trees undoubtedly observed the leaf fungus, and perhaps even suffered productivity losses without necessarily knowing the specific nature or etiology of the disease. The actual number of such owners was probably limited both in absolute terms and relative to the number of potential planters. Cultivated stands in Amazonia during the boom were usually small, scattered, apparently lacking in the environmental conditions needed to foster the severe infestations experienced later (Dean, 1987:62; Schurz *et al.,* 1925:98,100). Moreover, the experience of planters who suffered significant productivity losses in this highly competitive industry was likely to have been neither widely shared nor widely appreciated by foreign investors and local functionaries, who generally valued scientific knowledge over local folk experience. State subsidized efforts to promote Amazonian plantation development in the early 1910s did bring about a large expansion in planted area (Dean, 1987: 62), and yet they were not designed around or informed by knowledge of leaf blight. This fact further suggests that the fear of leaf disease was not the chief obstacle to earlier plantation development.

The Marxist view would suggest that the tight tapper-trader relation prevented the introduction of wage labor as well as the accumulation of capital necessary to develop rubber plantations. As indicated earlier, such an explanation rests on rather weak microeconomic foundations. Moreover, substantial surplus was available locally for investment; again, the historical record indicates that at least some estate holders did invest capital in the planting of rubber trees (Dean, 1987:45). Perhaps even more important, the labor argument is insufficient. Wage labor clearly was not the only path to plantation development. If the transition to wage labor was blocked (for whatever reason), what prevented investors from using other forms of labor contracts? Subsequent experience elsewhere, especially in southeast Asia, indicates that rubber plantations can be developed and worked not only by wage tapping but also by share tapping, contract tapping, and fixed rent tapping (see Chew, 1991). Such alternatives certainly had been tried for wild rubber extraction in the Amazon, and therefore were available at least potentially for application in local plantation development. With such alternate paths available, the transition to wage labor could not have been the primary impediment to plantation development in Amazonia.

A more straightforward explanation for thwarted rubber plantation development during the boom rests on three features of the industry itself. First, labor was fundamentally scarce in the basin for reasons discussed in Chapter Four. Second, the opportunity cost of capital was very high, because of the profitability of wild rubber extraction. And, third, the maturation time of planted rubber in the region was perceived to be relatively long. Planted rubber trees were believed by Amazonian rubber experts at the time to need 12 to 15 years before reaching maturity (see Akers, 1912:103), though experience in Asia would show that plantations could be become profitable after only seven years (see Lawrence, 1931:15–16). These three factors effectively combined to create an economic environment in which most holders of capital would be dissuaded from investing in plantation development. As we argue below, the long maturation time of planted rubber trees (perceived or otherwise) made labor scarcity and the opportunity cost of capital binding constraint and thus was the principal reason for failed plantation development.[21] Our argument begins with an examination of the wage requirements for plantation establishment.

In contemplating the prospect of investing in a rubber plantation in Amazonia, the prospective investor would face the difficult problem of labor scarcity. To attract, retain, and monitor workers (in effect, to induce them to work as laborers rather than tappers), the investor would have to pay a conspicuously high wage. Such a wage would have to be competitive not only with urban wages but also the expected return to independent rubber tapping. In addition, the wage would have to be sufficient to discourage workers from making illicit sales of rubber and from giving only minimal effort to their work. Adding an incentive such as claims on future production from the planted rubber trees would be minimally persuasive to tappers because of the long maturation time of the trees relative to other income opportunities and their needs. By financial standards during this expansive boom, even seven years would have been a considerable time to forego returns on initial capital investment. The potential cost of plantation development therefore was considerable, particularly when compared with plantations under way in south Asia where daily wages were about one-eighth the level of Amazonia, and potentially more lucrative labor opportunities, such as tapping wild hevea rubber, were not present to engender similar monitoring problems (Akers, 1912:102).

Even if the labor problem in Amazonia could have been surmounted, the lure of wild rubber extraction and trading, with their promise of a quick

---

[21] Some observers do indeed suggest that labor scarcity was an important impediment to plantation development, as others point to the high opportunity cost of capital (see Akers, 1912: 101–105; Bunker, 1984:1032 and 1985:68–69; and Weinstein, 1983b: 31–32). However, none have shown as we do here the crucial importance of the long maturation time of hevea rubber, where both labor and capital are already profitably engaged, in decisions whether or not to invest in plantations during the boom.

fortune from only a modest capital investment, sustained the high opportunity cost of capital. Plantation development in Amazonia would have represented a sizeable amount of capital tied up over a long term; consequently, to succeed, plantations had to assure returns far above those to be gained from wild rubber extraction. Few entrepreneurs could justify investing capital in plantations when much more immediate and substantial returns could be earned by making merchandise loans to traders, patrons, and tappers. As long as untapped wild rubber trees were accessible, surplus remained to be captured, and widespread plantation development in Amazonia would wait. To the prospective foreign investor, the choice was clear: invest in a plantation in Asia, where wild hevea rubber did not exist, labor was plentiful (or could be imported cheaply), and food was relatively inexpensive; or invest through existing channels of finance in wild rubber extraction in Amazonia.

Implicit to our explanation of thwarted plantation development is the most common model for the introduction of plantations—monocrop cultivation on rubber estates. Yet other, potentially more propitious settings existed in Amazonia for the transition to plantations. Incremental cultivation on agricultural estates along the Amazon River and the enrichment of rubber trails by tappers of wild rubber were probably more natural means to plantation development and may account for much planting prior to the introduction of state-sponsored incentives in the 1910s. A brief assessment of these two alternatives further illustrates why plantation development was stifled.

On the many agricultural estates that were founded along the Amazon River and its tributaries where little natural rubber grew, owners faced the same labor constraint as did rubber estate holders (that is, scarcity and high opportunity cost), though a somewhat different set of investment opportunities. With sufficient capital, the owner might consider replacing pasture or cacao plantations with planted rubber trees and presumably thus obtain higher returns from the land. Given the even higher and more immediate returns to wild rubber extraction, however, the owner would be more inclined to send accumulated surplus into the forest with traders, patrons, and tappers while maintaining secure returns from the sale of livestock and foodstuffs in urban markets.

Of course, some land owners may have sought to plant rubber on a small-scale and, depending on the outcome, decide whether to incrementally expand the area of rubber holdings. Reports do exist of sporadic attempts at establishing rubber plantations on agricultural estates (see Dean, 1987:45). However, incremental conversion of estate land to rubber would have been a protracted and rather dubious endeavor, given planted rubber's long maturation time and the uncertain outcome of planting. Prospective planters would need some assurance that such efforts would pay off, unless of course the owner could refer to promising results of state experimentation or depend upon substantial state subsidies. At the time, a planter would have expected to wait at least a decade to gauge the productivity of planted rubber trees, and if initial results were encouraging would then consider expanding his holdings with trees that would only come into production in another decade. Government research and development in the domestication of wild rubber and risk capital raised on the London stock exchange to finance rubber plantation companies were instrumental in the development of Asian plantations. In Amazonia, potential investors could not count on substantial state-sponsored research and development during the boom, and few individuals could afford to wait for the uncertain result of planted rubber.

The practice of independent tappers of adding to their natural holdings by sowing or planting along the rubber trails was probably much more common than the literature suggests. Rubber trees had been planted by indigenous peoples since pre-history in small clusters near settled sites in Amazonia, not only for the latex but because the seeds were an appreciated food (see Dean, 1987:45). Prior to the peak of the boom, one writer at Manaus observed, "[m]any rubber men in this region [Grão-Pará] are planting a few saplings: along their estradas [rubber trails]—say 20–30 each year—by way of experiment" (Garnier, 1902: 281). With such planting, a tapper could slowly build up the number of rubber trees on the estate, minimizing risks to density-dependent plant pathogens and effectively capitalizing the property by investing some of his labor in planting. With wild hevea occur naturally at low densities, typically fewer than 50 individual trees per square kilometer, tappers probably could have increased the relative density of planted trees on the estate substantially before the risk of leaf blight—which requires close proximity for transmission—would be significant. Once an essentially pure, extensive, closed-canopy of hevea was established, the estate would be at much greater risk of severe damage from leaf blight.

Despite the promise of higher returns from planted trees along existing rubber trails, tapper-planters faced problems similar to those of other potential planters. They, too, had to wait a long time to gauge their success, and the labor required to ensure seedling survival probably weighed heavily against the returns from ongoing wild rubber extraction. The seeds of the hevea tree are a preferred food source for many Amazonian rodents, particularly the paca (*Agouti paca*), and predation on cast or even planted seeds in areas with naturally high densities of rubber trees would be heavy. Seedlings would be more likely to survive but would require tappers to develop a nursery and protect each seedling once transplanted in the forest over several years. The labor time necessary for successful and significant enrichment of the tapper's rubber trails is expected therefore to be rather substantial. Moreover, tappers looked to immediate, substantial returns to fill their own seasonal need for provisions advanced on credit. Thus, tappers with accumulated capital were more likely to invest in becoming patrons or traders, than rubber planters. Furthermore, over time, as the forest was transformed into a higher-density rubber plantation, tapper-planters would have had to hire extra laborers to work the trees, and thus eventually would face the labor constraint themselves.

The factor that stifled broad and successful development of rubber plantations in Amazonia during the boom was not a lack of investment capital, a threat of destruction by a pathogen, or a tapper-trader alliance. Had massive state subsidies been introduced, and had they effectively transformed the Amazonian rain forest to rubber plantations prior to the advent of Asian competition in 1910, Amazonian plantations still probably would not have been economically viable. Plantation rubber from the Amazon would have been much more expensive than rubber from the Orient, because of the significant wage differential and the damage that would have been caused in the closed-canopy stands by the South American Leaf Blight. Such factors explain why plantations could not have been a success in Amazonia—indeed, such has been the bitter experience of companies attempting to develop rubber plantations in Amazonia since the 1930s. The real constraints were the very success of wild rubber extraction, the relative scarcity of labor, and the long maturation period of the planted rubber trees. These factors combined to discourage investment in plantations well before wage differentials or leaf blight could have limited plantation viability.

Another Perspective on Performance—Local Surplus Retention in the Wild Rubber Industry

We have suggested at several earlier points in this monograph that local surplus retention during the Rubber Boom was substantial and indeed sufficient to justify our shifting the focus of discussion away from the rubber sector's failure to generate surplus and towards the type of investment patterns that local accumulation engendered. Our argument rests on two important conclusions drawn from analyses above: (1) the Amazon rubber industry produced large resource rents; and, (2) a significant portion of those rents, due to the very organization of the rubber industry, was retained by local economic agents that included the state.[22] It seems worthwhile at this juncture to marshall our arguments and evidence for these two conclusions, before going on in Part II to investigate the patterns of accumulation and how private investment and state expenditures of such surplus, in turn, shaped the regional economy of Amazonia.

Until Asian plantation rubber swamped world markets in the 1910s, the Amazon was the world's preeminent supplier of rubber. A conservative estimate of the region's market share in the boom era would be 60 percent of the world supply. Two physical features of wild rubber suggest why substantial surplus was generated by its extraction. First, the native trees

---

22 We do not argue here, however, that locally retained surplus was sufficient to sustain economic development, another problematic counterfactual to construct, rather that surplus was retained and invested in a manner that engendered fragility. Leff (1973:687) makes a somewhat similar point when he compares the rubber and coffee booms in Brazil. He suggests that rapid growth in income associated with booming exports in rubber was not a sufficient condition for regional industrialization because the resulting market and income expansion may not have reached a sufficiently large scale to sustain industrialization.

that yielded rubber latex were highly dispersed along extensive and remote frontiers, which meant that opening new areas involved rapidly rising unit costs due to increased setup and transport costs as well as higher risks. Rapidly rising unit costs for new estates in turn meant the possibility of higher returns to prime sites. Second, the alternative to wild rubber extraction—plantation rubber—was thought to require more than a decade to establish. Thus, the inelastic supply of rubber combined with expanding world demand to generate large returns from wild rubber extraction.

Evidence abounds of the high returns engendered by wild rubber extraction in Amazonia. At the most abstract level, the massive influx of labor and capital over the course of the Amazon Rubber Boom and the dynamic extension of extraction to the furthest reaches of the basin strongly suggest the lure of higher than normal returns. In fact, financial returns to labor were among the highest levels in modern Amazonian history, driven by the lucrative nature of the extractive activity in the rubber sector. Amazonian states were able to levy sizable *ad valorem* taxes (10–20 percent) on rubber exports and even higher import duties on key consumer goods which generated massive revenues for public investment. Such levies and revenues could not have been sustained if alternate sources were not competitive enough to take advantage of the increased cost of Amazon wild rubber.

Perhaps the most compelling evidence of the accrual of large resource rents during the Amazon Rubber Boom is revealed by what did not happen to production levels in the Amazon during the price collapse of the 1910s. As prices plummeted following the flood of Asian plantation rubber, output from Amazonia, and the main producer, Brazil, did not fall notably. Between 1911 and 1917, annual wild rubber production from Brazil varied less than 10 percent of the output level in 1910, and production in 1912 was even higher than at the price peak; by 1918 and 1919, Brazilian wild rubber output accounted for 75 and 84 percent, respectively, of the 1910 production level (see Santos, 1980: 236) despite the 83 percent price decline over that period. Given that the technology of wild rubber extraction did not change notably during the period, and that ten years was more than sufficient time for agents to leave rubber extraction (if the industry had become a money-losing venture for most of its participants), the persistence of wild rubber production at boom era levels suggests the generous level of resource rents that must have been available to all but the most marginal of production locales in the region prior to the bust.

Our microeconomic analysis presented in Chapters Three and Four point to several explanations as to why local retention of resource rents and surplus was probably rather substantial. Decentralized extraction and transport operations, ease of entry at the various levels of the industry, the multitude of traders operating on the region's rivers, and the mobility of labor and product all meant that the potential for monopolizing or dominating any stage of the industry was minimal and was probably limited only to the remotest of rivers where rubber barons were able to rule for brief periods. The resulting competitive structure meant that foreign investors and large domestic concerns could not monopolize readily, as the United Fruit

Company did in bananas in Central America or the Aluminum Company of America did in bauxite in the Guianas around the turn of the century (Barham, nd., 1994). Most large-scale foreign investment in wild rubber production and trade came late in the boom, and was widely viewed as unsuccessful (Weinstein, 1983b:172–82).

Local surplus retention was by no means limited to the holders of wild rubber estates. Rubber trees constituted the relatively abundant factor in the Amazon basin, whereas labor and capital were more scarce. In general, both labor and capital sought high returns that were consistent with their scarcity, the mobility allowed by most river areas, and the risks involved in wild rubber extraction. In addition, the high cost of such a dispersed and decentralized extractive activity meant that tappers had certain advantages: they were both difficult to monitor, especially in terms of their rubber sales, and costly to replace should they fall to illness or desert their trails. These features suggest that tappers were able to capture on average, in what was a highly variable and risk-ridden environment, some portion of the rents embodied in wild rubber extraction.

Although good data on the actual average returns to tappers have yet to be found, substantial indirect evidence suggests that returns to tapping were attractive. The high urban and rural wages for unskilled labor during the peak boom years, one to two dollars per day, reflected the opportunity cost of rubber tapping. Moreover, some tappers became patrons and traders, capitalizing their way into a position of ownership in the industry (Weinstein, 1983b: 24). Semi-independent and independent tappers also cleared land, planted perennial crops, and built houses, often in the cities or towns near where they tapped, intending to occupy them in the off-season and perhaps move permanently to the cities after accumulating enough assets. Direct evidence of the accumulation patterns of patrons and traders is also lacking, but contemporaneous accounts suggest that their returns exceeded substantially those of tappers because of their additional access to other valuable resources, particularly capital.

In sum, if the performance of the wild rubber industry was to be assessed primarily on the basis of the quantity of capital accumulated and the propensity for local retention, the industry could only be seen as a singular success. The Rubber Boom brought several decades of unparalleled prosperity and growth, and, just as important, generated many of the preconditions considered necessary for broad, sustained regional development: Amazonia became integrated into the world economy; trading networks extended into the remotest reaches of the basin; massive port facilities were created; labor-intensive economic activities were pursued; large urban centers with vigorous consumer demand sprung up along the Amazon River; and, abundant surplus was available for investment that could have fueled development in other sectors. Yet, we know that the boom failed to lead to broader and long-term sustained economic development in Amazonia. Clearly, no assessment of the rubber sector's apparent weaknesses—whether they be thwarted plantation development, resistant social relations of extraction or the unresponsive supply of wild rubber—is sufficient to explain the broader outcome of unrealized, multisectoral economic development following decades of dramatic expansion associated with the boom. Why the boom failed to propel the region onto a more successful, long term development path lies not therefore with the performance of the wild rubber industry, but rather in how rubber wealth shaped the emergent political economy of Amazonia.

Chapter 6

TAKING STOCK: LESSONS AND CHALLENGES OF THE WILD RUBBER INDUSTRY

Introduction

In the previous three chapters, we examined in some depth the industrial organization and microeconomics of wild rubber extraction during the Amazon Rubber Boom. Part II of the monograph, beginning with Chapter Seven, integrates our key findings about the wild rubber industry's organization and performance into an account of how the Amazon Rubber Boom shaped the development of the regional economy. The present chapter serves as a bridge between Parts I and II, by bringing forward the key empirical findings of our analysis. Moreover, the chapter offers the opportunity for the reader (and the authors) to reflect upon our findings so far.

In developing our main arguments regarding the wild rubber industry's organization and performance, a number of important empirical issues and challenges arise. Although we thoroughly examined and carefully marshalled evidence from many primary and secondary sources of historical information, we encountered a notable paucity of data on certain important points that highlights the need for further research, particularly archival work. More detail is needed, as discussed below, on the temporal and spatial variations in industry outcomes, if the specific local effects of the sector on the development processes in various parts of the region are to be better understood. In a sense, many of our arguments were painted with a rather broad empirical brush. We have drawn our evidence with the intention of making a credible claim or revealing some underlying logic, rather than to provide comprehensive, fine-grained accounts of the local histories of the boom. That task remains a vital one.

We pause here also to reflect on some of the conceptual lessons related to the methods employed for studying the organization and performance of modern resource extractive industries. Among our original motivations for undertaking this work was a sense that an improved understanding of the wild rubber industry would equip us with the insights and tools needed for more advanced study of resource extraction today in Amazonia and elsewhere in the developing world. In retrospect, we are even more convinced of the value of such historical work. We hope to persuade the reader on this point by linking the conceptual advances made here with the challenges of understanding posed by modern resource extractive industries.

We open the chapter by drawing forward from previous chapters those empirical lessons that are central to a revised understanding of the boom's regional economic development outcomes that we develop later. We then revisit the modified Structure-Conduct-Performance Model, as discussed in Chapter Two, and assess more explicitly how the model informed our inquiry. Subsequently, we consider the benefits of explicitly integrating risk and transaction cost issues into the microeconomic account of social or contractual relations in resource extractive economies. Finally, we raise the important question of how the microeconomic study of an extractive sector might differ under non-boom conditions (as prevail in many areas of Amazonia, today), and what other factors must be incorporated in the analysis that were more readily omitted in the study of a boom sector.

Empirical Lessons from Amazonia's Wild Rubber Industry

We began our analysis of Amazonia's wild rubber industry by signaling the region's position in the international market, particularly with respect to alternate sources of supply and the nature of demand during the boom era. Amazonian wild rubber played a premier role in world markets for several decades, because the region provided the finest quality rubber in the largest volumes to the burgeoning industrial economies of the U.K, U.S., and Europe. The fact that wild rubber was naturally dispersed in the rain forest and available from innumerable locales all over a vast region made entry barriers of any sort to extraction and trade difficult to create and sustain. Although large fluctuations in the market price of wild rubber introduced considerable risk to investors, especially in an era when options for hedging and futures' trading were limited, compelling evidence is available that exceptional returns were made by many of those parties directly involved with rubber extraction and trade.

The "high return" nature of the wild rubber sector can be characterized perhaps most parsimoniously as follows. World supply during the boom years was rather inelastic, because of the highly dispersed nature of wild rubber collection and the lack of other prime sources like the Amazon. Therefore, many areas in the Amazon basin—those which were not too remote or had high yielding rubber trees—became sources of wild rubber at costs that were well below the world price, what economists often call "infra-marginal sites." For participants in such prime locations, the potential to capture significant economic rents when prices were high must have been quite great. As prices rose strongly for much of the last two decades of the boom, producers at such infra-marginal sites probably realized high returns over an extended period.

Although an extractive activity, such as wild rubber extraction, may bear the promise of high returns during a resource boom, participants must first face certain risks and transaction costs inherent to engaging in the activity. Indeed, the rubber industry, particularly in the collection and trade in wild rubber, was distinguished by the prevalence of high risks and transaction costs. This basic characteristic made certain types of contracts or exchanges among agents (e.g., wage labor) rather problematic while making other types of contracts relatively attractive (e.g., piece rate) by terms of the incentives they offered participants.

Contractual arrangements were conditioned not only by risk and transaction costs, but also by the endowments of individual agents and the relative abundance of the key factors of production. In particular, most tappers lacked sufficient capital to provision themselves with the food and other basic working capital for a full season of tapping. At the same time, tappers saw that taking too much time off from tapping to cultivate subsistence crops or fish/hunt would not be profitable, especially when the relative returns to rubber tapping were so high and limited to the same season of the year as many subsistence agricultural activities. Credit in the form of provisions was, therefore, essential to these tappers as a means to concentrate their labor on the high return activity. This need for advance provisioning meant that in order for wild rubber to be extracted on any scale, some sort of credit market would have to evolve that would serve tappers even in the remotest of locales.

The debt-merchandise contractual relation between traders, patrons and tappers became the central means of providing credit to low wealth tappers and, simultaneously, of securing scarce labor to work hevea rubber and to deliver relatively reliable supplies of rubber. Much like the contract farming example presented in Chapter Two, monitoring problems associated with other forms of labor and credit relations were mitigated through the debt-merchandise arrangement. On the credit side, tappers obtained a steady source of advance provisioning, which allowed them to pursue tapping full-time, and to ride out difficult times, such as illness or low price periods, by extending their credit line with the parton or trader. At the same time, the payment of a piece-rate on the rubber by traders or patrons provided tappers the incentives to actually deliver their rubber, with the potential difficulty being whether the tapper felt he was receiving a fair price from the trader. The presence of itinerant river traders was one source of price discipline on the debt-merchandise traders. Another was the opportunity that tappers had in their solitary work to set aside (i.e., hide) some of the crude rubber for later shipment or sale.

As argued above, the dynamic incentives of the debt-merchandise contract were mutually reinforcing and suggestive of why labor-capital relations might have been so "durable" and "resistant." The tapper gained insurance from cultivating good relations with the trader or patron, while the trader and patron secured a steady source of labor (thus avoiding the transaction costs of replacing an exiting tapper) by developing good relations with the tapper. Moreover, for the patron, such a relationship would improve the likelihood that his hevea trees would be tapped in a sustainable manner. Thus, such concerns—along with the overall scarcity of labor, the mobility provided by river travel, and the difficulty of monitoring workers in such a dispersed and decentralized activity—served to limit the power of creditors (i.e., patrons or traders) and helped to balance the bargaining table for tappers.

At the same time, the high risk features of wild rubber collection and exchange created the need for significant mark-ups along the chain of trade. Although certainly in some (and perhaps many) instances, the pyramid structure of pricing in the trade system may have engendered "exploitative" bargaining (because a particular agent in the chain was, in fact, limited in mobility or ignorant of broader market conditions), the pyramid scheme was not inherently unfair, or *prima facie* evidence of exploitation. That determination could only be made in the specific context of the exchange. Where tappers were relatively mobile, where itinerant traders plied their wares on the river, striking deals with whomever would trade wild rubber or other goods, where tappers owned their own estates, then it was expected risk—not monopoly relations—that sustained the pricing pyramid of wild rubber on the river. Conversely, where such conditions did not hold, and tappers were in a weak bargaining position, then the margins in the pricing pyramid probably would have been larger. The empirical challenge of discerning competitive from exploitative outcomes lies in distinguishing between the roles of risk and bargaining power in determining price margins, in specific locations or situations.

In the case of caucho, where the trees were felled for their rubber, a greater variety of contractual arrangements arose than in hevea tapping. Expeditions were mounted by compact teams of men searching the vast interfluvial forest for such trees. The one-shot nature of caucho gathering meant that such teams formed for a limited period; thus, no need existed to build in "reinforcing" relations to reduce longer term transaction costs. Moreover, team work served to lower enforcement costs. Perhaps not surprisingly, some of the most hierarchical or coercive capital-labor relations evident in the boom arose around caucho extraction in highly isolated locales, where enterprising rubber barons contracted with local Amerindian groups to extract wild rubber. Such certainly was the case in Arana's domination of the Putumayo River of Peru.

The principal lesson to be drawn about the performance of the wild rubber industry is that given:

- the high returns to Amazonian rubber on the world market;
- the general scarcity of labor and capital in the region and abundance of rubber trees throughout much of the boom; and,
- the risk and transaction costs which framed contractual relations of extraction and trade;

significant economic returns (i.e., a major share of the rents involved), accrued during the boom to the scarce factors of labor and capital. As mentioned above, this broad generalization would be modified according to a number of specific conditions (e.g., the locational setting, the timing of entry into the sector, the initial endowment of the economic agent, misfortune along the way, etc.). Nevertheless, compelling data ranging from the wage levels reported at the time for unskilled labor in the port-cities to the volume of imports entering the region long after the boom had ended suggest the capture of substantial returns by local agents.

In our discussion of problems of the wild rubber industry, we re-examined and critiqued the purported performance shortfalls of the sector. Of the three performance problems examined, the most central one was certainly the issue of thwarted plantation development. Why was more intensive plantation production of wild rubber not achieved in Amazonia, despite what would seem to be conducive conditions (e.g., abundant hevea seed stock, the infusion of considerable foreign capital to the region, the competitive imperative of Asian plantation rubber, and private and state initiatives targeted at that transformation)? The key to our alternative explanation takes us back to the very success of the wild rubber sector given the relative endowments of trees, capital, and labor in the region and the high returns that could be earned by participants in tapping and trading in wild rubber. It was the profitability of wild rubber in Amazonia, not South American Leaf Blight, that kept investors from pursuing plantations on the massive scale that was realized in south Asia. Even if the blight could have been overcome, plantation development in Amazonia would have not have been competitive with Asian plantations because of the shortage of labor and its associated opportunity costs, the high cost of food provision for labor that might have been imported into the region (under various immigration schemes), and the difficulty of holding onto any form of "indentured labor supply" in the riverine-jungle environment. This brings us to our primary conclusion from our analyses in Part I: the explanation for the major economic depression in Amazonia that followed the steep decline in rubber prices in the 1910s clearly does not lie in the performance shortcomings of the wild rubber industry itself. This finding suggests that our attention should focus elsewhere, specifically on how the rubber sector shaped the overall economy of the region.

Future Research on the Organization and Performance of the Wild Rubber Sector

Research efforts that attempt to approach a long-standing problem from a new direction have the potential of raising more questions than they answer. Phenomena that had been unobserved, ignored, or simply passed over by previous researchers become bathed in the light of fresh inquiry and call out for explanation. Many of these phenomena, however, are likely to remain only partially explored, because in laying forth their new approach, researchers are likely to face more questions than they can possibly answer, especially if the attendant conceptual and empirical exigencies are large. As a result, researchers are faced with the task of making strategic choices about which questions or phenomena to tackle first and which to leave for further inquiry.

Our strategic choices were shaped by the following goals. First, we wanted to provide a more careful microeconomic treatment of the rubber sector than had been done previously. Second, we sought an approach to industry analysis that would be flexible enough to account for the significant spatial and temporal variations we encountered in our reading of accounts of boom. And third, we were drawn to the larger question of why the boom did not give rise to more sustained economic development in Amazonia. Given our third goal, at a certain stage in our research, we had to lay aside a number of important issues and proceed to addressing the problem of the boom's role in shaping the regional economy. We raise these issues further below.

As important as the outstanding questions, however, were the impediments we encountered in undertaking our analysis. The most fundamental impediment we met was the lack of precise microeconomic information on the costs and returns of rubber extraction and trade to various participants as well as on the terms of debt-merchandise contracts, specified for particular places and times. In the few studies where estimates are provided of prices paid for rubber and charged for the merchandise inputs, we could not determine with confidence whether such prices were clearly those paid by the trader to the patron or to the tapper, and vice versa. Where the risk of tapper attrition is high, such a distinction is decisive: the patron would need a significant price wedge as compensation for being the tapper's creditor, given the riskiness of the venture and the limited options for local diversification. Furthermore, extant estimates of the prices charged and received are not accompanied by such essential information as the variability of returns to the concerned parties or local attrition rates of tappers. Returns variability is crucial in understanding the distribution of rents and risks to the parties involved and hence the competitive nature of the contractual arrangements.

The type of data necessary to allow for a careful exploration of the debt-merchandise relations would be found in old accounting records kept by trading houses and patrons. Although we are hopeful that such records may be held in regional public archives or private collections, we found no secondary analyses employing such data that confirmed this possibility. Given the paucity of primary data available to us, we sought out for our analysis other types of information that would frame the terms of the debt-merchandise arrangement and the returns offered to the various participants. Our findings suggest certain ranges of possibilities in returns, as well as the attendant geographic and economic factors that influenced returns across space and time. The basic question, however, of specifically how returns and risks were distributed among traders, patrons, and tappers in different locales and periods remains open to scholars for future study.

Those factors that influenced financial returns and risk in the industry also merit further inquiry. Of particular interest would be the contrast between rubber estates along rivers that were closed to trade and those along "open" rivers. Where trade along a river was limited by geographic isolation (or limited entry and exit points) and the "strong arm" of a rubber baron, then the returns and risks might have been biased against local tappers, especially if most were new to the area, with little knowledge of the locale and the relative terms to be offered. A nearby river that was open for trade (or only briefly closed) would provide an interesting opportunity to compare risks and returns to participants, the supply conditions (both levels and productivity) on these rivers, and contrasting local development impacts and legacies. A map identifying which Amazonian rivers and tributaries were actually open/closed to trade (and for which periods) is needed to guide researchers in the selection of such cases.

In addition, research is needed to better understand the role of the initial endowments of tappers in explaining the variations in sectoral outcomes. Certain tappers clearly were better positioned to capture a higher share of returns to their labor than others. Some tappers, for example, had extensive knowledge of the forest or the rubber trade; others held squatting rights to their rubber estradas; and some possessed an initial stock of capital. Such endowments most likely were to be found among people who had been making a living from Amazon forests and rivers prior to the boom (i.e., the caboclos or ribereños), and among certain immigrant groups from other regions as well. In contrast, the endowment base of poor Cearense immigrants would have been much more limited, and therefore limiting. Comparison of the return-risk experiences of these different groups of tappers (in proximate locales) would reveal much about the nature of debt-merchandise contractual relations, the potential of different groups for economic advancement, and about the varied locational impacts of the boom both within and outside the Amazon basin. On the later point, one could compare, for example, the effects of the boom on two towns in northeastern Brazil who sent immigrants to Amazonia at distinctive historical moments as a means of better understanding outcomes for rubber tappers and the legacies of the era. In addition, a comparative case study could contrast the experience of tappers on "upriver" and "downriver" estates, or between hevea tappers and caucho gatherers.

The returns and risks issue also could be explored fruitfully by considering the extent of upward socio-economic mobility among the various participants of the rubber industry. To what extent did tappers in certain areas accumulate sufficient capital to become patrons or traders? Conversely, did some patrons and traders also tap rubber as a sideline activity or after suffering losses in their other roles? What about the children of tappers? Were they able to pursue educational or other opportunities in the urban areas that developed during the boom era? To answer such questions, researchers would need to construct detailed local histories of estates, communities and families that had been involved in the Rubber Boom. In the case of the Tahuayo River, near Iquitos, Peru, where historical research has been pursued on resource use dating back to the boom era, Coomes (1995) has found considerable anecdotal evidence that some tappers had, indeed, been able to accumulate tangible wealth (e.g., rubber estates or urban real estate) and to invest in the advancement of their children, or relatives, into urban professions and businesses.

Although we are convinced that the purported failings of the industry do not adequately explain why the boom's promise of prosperity was not fulfilled, one performance issue does merit further study—the plantation development experience in Amazonia. Of particular value would be studies of specific attempts to establish plantations *prior* to the 1910s.[23] Such studies would allow a more empirically informed comparison of the evidence supporting Dean's thesis (i.e., that South American Leaf Blight was the primary obstacle to plantation development) with our argument that other fundamental economic problems (e.g., the costs of securing the labor effort needed, either through higher wages or monitoring, or the high opportunity cost on financial capital involved in wild rubber extraction), would have thwarted the effort well before serious leaf blight problems emerged. Archival and financial records on this issue may offer more promise for in-depth study than on others issues discussed so far. Such research would also provide further evidence on the type of returns received by tappers by showing what labor costs investors faced in their plantation efforts.

Conceptual Innovations in the Study of Forest-Based Extractive Industries

In Part I, we used our industrial organization analysis to frame a careful look at how previous writers had evaluated the impact of the wild rubber sector on Amazonian development and to reassess their conclusions. We found that most of the previous interpretations which emphasized performance problems in the sector did not hold up to the findings of a more careful microeconomic analysis. At the heart of the growing popular and scholarly literature on *current* forest extractive activities in the Amazon basin, we note a similar (though perhaps more vigorously expressed) concern with conservation and development impacts. We believe that such concern could be acted upon most effectively if researchers and conservationists recognize the potential value of carefully structured, microeconomic analyses of extractive industries. Getting the microeconomics of natural resource extraction right for effective conservation and development is perhaps the

23 We stipulate the time prior to 1910 because once the Asian plantations came fully into production during the 1910s and prices of world rubber began to fall precipitously, export-oriented rubber plantations in the Amazon would have been uncompetitive with Asia because of higher labor costs. Such cost differential did not, however, dissuade the state (at least in Brazil) from undertaking to promote rubber tree planting in Amazonia during the 1910s.

single most important challenge for future work on extractive industries in fragile environments.

To date, surprisingly little empirical research has been done on the microeconomics of traditional resource extraction by Amazonian forest peoples. Until recently, forest product extraction and swidden-fallow agriculture had been seen to offer forest peoples little more than perennial poverty and exploitation (e.g., see Wagely, 1953:100; Furtado, 1963:147; Alvim, 1979:22; and Sánchez, 1981:382). However, recognition of the rich biodiversity and underlying fragility of rain forest environments during the late 1970s and early 1980s gave rise to an explosion of interest in the traditional ways and knowledge of Amazonian tribal and peasant peoples. Such interest is reflected in the numerous ethnographically-styled studies among forest peoples of Amazonia that show the ecological and subsistence benefits of traditional practices (e.g., Parker *et al.*, 1983; Hiraoka, 1985 a,b,c; Posey and Balée, 1989; Pinedo-Vásquez *et al.*, 1990). Moreover, recent studies also suggest that traditional Amazonians do indeed derive significant economic (cash) income from relatively sustainable use of the forest (see Padoch *et al.*, 1985; Padoch, 1988; Anderson, May, and Balick, 1991; Redford and Padoch, 1992). Accounts are now available for gatherers of açai and babassu in Brazil near Belém, rubber tappers of Acre, and ribereños near Iquitos, Peru.[24]

Such studies represent "first attempts" at gauging the approximate economic returns to various activities, ranging from agroforestry to forest product extraction. Most studies unfortunately cannot claim statistical representativeness, rather only that results are indicative of potential returns, because these studies rely heavily on small, convenience-selected samples and or anecdotal information on yields, prices, and so forth. Moreover, they tend to employ dubious economic valuation methods, and give only sparse attention to the actual economic decisions made by households participating in these activities. As a result, a "second wave" of more in-depth and exhaustive studies on household income generation and expenditures are currently being undertaken by researchers and conservation organizations working in Amazonia and elsewhere in tropical Latin America. Such studies promise to provide a firmer foundation for rain forest conservation and development initiatives.

The conceptual lessons from our study of the wild rubber industry are applicable in several ways to current research efforts on extractive industries in forest environments. Our central conceptual offering to the current literature is the systematic approach offered by the modified Structure-

[24] For example, Hecht, Anderson, and May (1988), Anderson and Ioris (1992), Anderson and Jardim (1989), Schwartzman (1989), Padoch *et al.* (1985), Padoch (1988), Anderson, May, and Balick (1991) and Chibnik (1994).

Conduct-Performance (SCP) Model we used. A similar approach to the study of current extractive industries would advance research efforts by:

- *Promoting a cleaner separation between the conduct of industry participants and the resulting performance outcomes.* As an example, complex contractual relations and/or governance structures, such as the debt-merchandise contract or the foreign company, often are taken as emblematic of a means to exploit local peoples, when in fact the contractual relations among different participants actually may have emerged for other reasons (e.g., transaction costs, risks, and the distribution of key endowments). This effort to distinguish between conduct (i.e., strategies and contractual relations) and performance (i.e., efficiency, returns, and risk burdens) is not to deny the potential for one class or group of industry participants to gain the upper hand in certain transactions; rather to ensure that the source(s) of that upper hand can be explicitly identified so as to avoid remediation efforts targeted at the wrong problem.
- *Highlighting the importance of risk and transaction cost in shaping contractual and other institutional arrangements between economic agents.* Omission of such key variables in models aimed at explaining extractive resource use or the organization of extractive industries can result in other variables being falsely identified as playing a significant role. In the absence of an appreciation of the microfoundations of contracts in extractive activities, explanations of price markups in the trade often point instead to price gouging, and giving rise to policies aimed wrongly at eliminating intermediaries. This problem is much more likely to arise when an explicit and comprehensive explanatory model is not set forth to guide analysis.
- *Focusing attention on the role of extractive product characteristics and the geographic and technical conditions surrounding its extraction and transport.* Failure to ground the analysis solidly in these dimensions raises the likelihood that important variations are overlooked in industry conduct, structure, and performance because of the distinctive features of an extractive product and the distinctive locational and environmental context in which the extractive activity proceeds. Resource extraction is a geographically constituted economic activity.

Particular attention in the application of the modified SCP approach must be given to the position of the extractive industry in the regional economy. Wild rubber was unquestionably the dominate economic activity in Amazonia and responsible for driving the boom. Where another product of similar importance is to be studied, then our analytical focus on structure, conduct and performance will be useful. However, where a resource is not so prime, profitable or predominate—as is the case for most extractive resources today in Amazonia—the problem of understanding the organization and development outcomes of the industry becomes a more complex and qualitatively different challenge. In such a case, emphasis must shift to understanding the place of that product in the region economy, the alternate economic options and returns that are available, and the underlying logic of decisions by local households and other economic agents to participate in extractive activities.

In large part, the "second wave" of research on extractive activities among forest peoples is currently facing this challenge. Today, there are very few instances of major forest-based extractive resource booms. Instead, forest product extraction generally is being pursued by rain forest peasants (indigenous or otherwise) as part of a diverse portfolio of household agricultural and extractive activities. Conceptual guidance as to how to structure an analysis of such activities is available from economic studies of peasant agriculture in developing countries, though important modifications would have to be made for "extractive" activities because of their distinctive character. Some of the key questions to be asked, where a resource industry is not "booming" or dominating returns to other sectors, include:

- *What factors influence the participation of forest peasant household participation in extractive activities?* Whereas fishing, hunting and non-timber forest product extraction represent major sources of cash income for some households near prime forest resources, others in the same area live primarily from agriculture (see Coomes, nd). Why then do some households rely more heavily upon the forest than others? What factors influence livelihood choice through time? Under what circumstances and conditions do households specialize in certain extractive products?

- *What types of contractual or other institutional arrangements are employed to engage peasant households in extractive activities?* In Amazonia today, natural resources are extracted under a wide range of agreements struck between collectors and traders or buyers. The specifics of such agreements are important because they shape financial returns to households and reflect the key constraints, risks and transaction costs entailed by particular extractive activities. How do the specific characteristics of the product and the collection process influence the types of arrangements that are possible? What are the key risks and transaction costs? How do endowments such as equipment, environmental knowledge and extractive skill influence the bargaining position of the household? How does the "agriculture option" where land is abundant influence household bargaining power?

- *What role does resource extraction play in wealth accumulation and investment strategies among peasant households?* Much of the focus of integrated conservation-development initiatives lies in enhancing rural incomes from extractive activities, yet wealth and investment may be more relevant factors in terms of longer term resource use behavior and household welfare outcomes. Under what conditions is resource extraction a more promising path to accumulation than other activities (e.g., agriculture, wage work, etc.)? Is resource extraction a stepping stone toward investments in potentially more resource-conserving and welfare-improving activities, such as intensive floodplain agriculture, agroforestry, or education of children? If so, under what conditions? If not, why not? As households accumulate wealth through resource extraction, do they adopt or invest in extractive technologies which are higher yielding and more destructive to key resource stocks? If so, under what conditions, and why?

- *How can resource extraction activity be managed or modified to be more sustainable?* Whereas low levels of resource extraction near prime forest areas are more likely to be more sustainable, the management challenge of conservation and development initiatives intensifies when demand for extractive goods rise and/or few alternate income opportunities exists. Are there identifiable substitute activities or refinements in current resource extractive practices that can both raise incomes and improve resource conservation? What are the main impediments to more sustainable resource use?

In those cases, however, where the extracted resource is sufficiently prime in terms of demand and competitive supply sources to generate much higher returns than alternative activities, then the study can be decidedly more focused on the industrial organization of the dominant sector. Our work on wild rubber during the boom had the luxury of that option. Indeed, financial returns were so high that tens of thousands of new participants in the industry and its ancillary activities were drawn to the region during that era. This feature allows less focus on participation decisions of economic agents (which are not really in question) and more on the industrial organization of the specific extractive activity. In addition, the dynamics of the boom sector and how that sector influenced the evolution of the regional economy become particularly important to understand, as we shall now see in Part II for the case of wild rubber.

PART II

✢ ✢ ✢

Investment, the State and Dutch Disease in the Amazon Rubber Economy

✢ ✢ ✢

Chapter 7

THEORIES OF AMAZONIAN UNDERDEVELOPMENT

Introduction

Few periods in South American history have so captured the imagination and begged the attention of scholars as the Amazon Rubber Boom. For fifty years, the extraction of wild rubber from the jungles of the Amazon fueled unprecedented economic expansion in the region; per capita incomes in the Brazilian Amazon climbed by 800 percent; the regional population increased by more than 400 percent; urban centers and secondary towns blossomed along the river banks; and the vast Amazonian forest lands were integrated into national political spheres and the international market economy.[25] Amazonia became a major trading region, with one of the highest trade coefficients in the world.[26] But when low-cost rubber from British plantations in Asia flooded world markets in the 1910s, rubber prices plummeted, sharply curtailing financial returns from wild rubber extraction in the Amazon. The price shock drove scores of traders and export houses into bankruptcy when they were unable to collect debts that were based on the future value of rubber. Urban real estate prices in Amazonian cities crashed, and service industries withered along with their customers' incomes. By the early 1920s, the boom was over, and per capita incomes had shrunk to pre-boom levels. Today, nearly a century later, such incomes (in real terms) have yet to return to boom levels in many areas despite massive state investment in Amazonia.

---

25 For estimates in growth in population and gross domestic product between 1800 and 1970 in the Brazilian Amazon, see Santos (1980:12–13).

26 The extent to which the Amazon had become a major trade region is evident in the data presented by LeCointe (1922:I:283–84). In 1912, the Brazilian Amazon had the eighth-highest trade coefficient (that is, the value of exports plus imports per capita) in world (459 francs, or $89 per capita), less than Holland ($346) or England ($113) but greater than Germany ($83), France ($56), the United States ($26), and Brazil as a whole ($19).

Recent scholarly contributions from North and South suggest a florescence of academic interest in the era, with more than a score of books and articles published during the 1980s. Most of these works pursue the longstanding issue of why this period of massive growth and prosperity did not lead to sustained development in Amazonia. This question is particularly appropriate today as the governments of Amazonian countries look to the region as a trove of untold wealth, a vent for surplus, and/or a frontier for absorbing the landless poor, and begin to consider the challenge of how best to promote more economically and environmentally sustainable development in the region.

In Part I, we set the stage for our reinterpretation of the Amazon Rubber Boom by examining the industrial organization and performance of the wild rubber sector. Attention was directed away from the purported performance failures of surplus generation and efficiency in the wild rubber sector and toward the broader patterns of accumulation and development that occurred in Amazonia during the boom era. The principal aim of Part II is to explore the development patterns of the regional economy and to identify the development legacies and lessons of the era. At the heart of the account is the analysis of private and public investment and spending patterns, and the distorted economic structure that resulted. In this distorted structure and the attendant vulnerability to weakness in rubber prices, we find the reasons why the promise of long-term economic development in Amazonia was unfulfilled.

Part II is organized as follows. In the present chapter, we review and critique the prevailing general theories of why the Rubber Boom failed to provide for sustained economic development, all of which point to some degree to the lack of local surplus retention as the principal cause. We then offer an alternate theory of distorted development, based on a macroeconomic model known as the "Dutch Disease" model. Formalized by Max Corden and Peter Neary (1982), the model is used increasingly by trade and macroeconomists as a means of understanding the distortionary development episodes that can accompany resource booms in less developed economies, where the reliance on a single primary product export is frequently high.[27] Finally in this chapter, we briefly explore the types of information that are available on investment during the Rubber Boom and on the role of the state.

In Chapter Eight, we explore the patterns of private investment during the Amazon Rubber Boom and document the evidence of Dutch Disease effects in the Amazonian economy during the boom. Because Amazonia was, in large part, integrated into the world economy by the Rubber Boom the degree to which the economic structure of the region was completely shaped by the boom is perhaps unparalleled in other recent episodes

---

27 A recently published synthesis volume of a World Bank study on the macroeconomic experience of developing countries includes several instances of Dutch Disease outcomes (see Little *et al.,* 1993). See also Auty's (1993) study of this "resource curse" among mineral-based economies.

elsewhere in the world. This fact may account, in part, for why the bust that followed the boom was so very dramatic in Amazonia, compared to the experience of other regions once their respective booms were over. Our attention in Chapter Nine turns to the role of the state in the Amazon Rubber Boom. We draw on evidence from national and state governments of Brazil, Peru, and Bolivia to portray and better understand the patterns of state taxation and expenditures during the boom. We show that the state acted through fiscal policies and other measures, to strongly reinforce private investment patterns which concentrated resources in the booming and related sectors.

We present in Chapter Ten our revised understanding of the structurally distorted nature of the Amazonian boom economy and its massive contraction in the 1910s. The chapter also offers a discussion of the development legacies of the Rubber Boom and points to a set of outstanding research questions to guide further inquiry on the macroeconomic outcomes of the boom. We conclude the monograph with a discussion in Chapter Eleven of the lessons offered by our analysis for better understanding the role of natural resource extraction in economic development. We suggest the vital importance and need for an integrated framework that explicitly links the microeconomic organization of the resource extractive sector with macroeconomic structure and performance outcomes. In this way, our book ends with a look ahead.

Prevailing Views on the Rubber Boom and Amazonian (Under)Development

Most contemporary work on the Amazon Rubber Boom examines the question of why the boom did not lead to sustained economic growth and social change in one or more of the Amazonian countries. From a review of literature on the boom, we identify three basic perspectives on this question and refer to them as: the dependency view, the political ecological thesis, and the Marxian view. For each perspective, we review the basic argument and draw on the findings from Part I to refute the contending explanation for why the Amazon Rubber Boom did not lead to sustained development.

Dependency and Underdevelopment

The dependency school provides a macro-level perspective on the problem of limited, sustained development in the wake of the Rubber Boom and is most evident in accounts of the era for Peru (e.g., San Román, 1975; Bonilla, 1977; Chirif and Mora, 1980; Flores Marín, 1987). According to this view, exceptional profits that accrued from the rubber trade were transferred out of the region and thus made unavailable for local development. Surplus was extracted within the region through unequal exchange maintained by debt-peonage and coercion. Foreign firms, perceived as operating as a monopoly and/or monopsony[28] (Bonilla, 1977; Flores Marín, 1987), and domestic elites (Santos, 1980; Haring, 1986) extracted the surplus and chose not to invest in the region. As rubber extraction and trade did not require major investments in local input or processing facilities, the Amazon basin is said to have been converted into an extractive enclave analogous to the coastal guano and nitrate rich areas of Peru (see Levin, 1960).

The argument that surplus transfer out of the region during the boom caused regional underdevelopment is singularly unpersuasive in this instance. High labor and local capital earnings and the existence of a variety of labor relations other than debt peonage raise serious questions about the effectiveness of unequal exchange as a mechanism for extracting surplus from the region. The many firms and markets involved in rubber suggest that trade was not monopolized by foreign (or domestic) firms. Moreover, direct foreign investment came late to the region, and was decidedly unsuccessful (Weinstein, 1983b). More important, as we document more fully in Chapters Eight and Nine, considerable surplus was retained in the region via private investment and state expenditures of export and import tax revenues. Sufficient evidence exists to suggest that the dependency interpretation is overgeneralized, simplistic, and incompatible with historical data from the period.

THE POLITICAL ECOLOGY OF UNDERDEVELOPMENT

The persistent leading role of extractive industries in Amazonian economic history and the concomitant degradation of rain forest resources inspires a second view that perceives the main consequences of extraction, including that of wild rubber, to be the underdevelopment of Amazonia and marginalization of its rural people (see Ross, 1978; Bunker, 1984, 1985; Domínguez and Gómez, 1990). This view is neatly summed up by the authors of an historical study of resource extraction in the Colombian Amazon (1850–1930) as

> ... al finalizar cada ciclo el saldo geográfico y social para la región era negativo: se había retrocedido en lugar de avancar ... la economía extractivista, como tal, no produce desarrollo. (Domínguez and Gómez, 1990:260)

---

[28] The distinction between monopoly and monopsony is simply this. In the former, a firm or economic agent is effectively the sole supplier of a particular good or service to consumers, and can thus price at higher, non-competitive, levels. In monopsony, a firm or economic agent is effectively the sole purchaser of a particular good or service, and can thus demand that the price be lower than it would be in a market with many buyers. A debt-merchandise trader could be viewed as both a monopolist and monopsonist if he is believed to be the exclusive seller of provisions to the tapper and the exclusive buyer of the tapper's rubber. Explicitly or implicitly, this would appear to be the view of the debt-merchandise relation that prevails in the dependency literature.

The political ecological thesis of underdevelopment for Amazonia is perhaps best elaborated by Bunker (1984, 1985). He argues that the extraction of natural resources from the Amazonian rain forests has produced a net out-flow of value and energy from the basin which reduces future options for productive (as well as extractive) activities in the basin. The continual dependence of local elites upon resource extraction to satisfy external sources of demand, shaped a political economy that reinforced the importance of extraction while undermining the resource base upon which the region depends. The result in Amazonia has been perpetual cycles of natural resource extraction, environmental destruction, impoverishment, and underdevelopment (Bunker, 1985).

The underdevelopment interpretation has a certain intuitive appeal. Anyone who has witnessed firsthand the predatory extraction of one plant or animal species after another from the rain forest and waters of the basin, or is familiar with the environmental history of Amazonia, can appreciate the concern over resource degradation. The political ecological thesis is limited, however, in its ability to reveal the logic of the rubber era or explain why sustained economic development did not follow the boom. The thesis is neither product-specific nor era-specific but rather treats extraction as a cyclical, long-term process of resource exploitation.[29]

Our critique of the political ecology thesis draws on a critical and pithy commentary by Martin Katzman (1987). He argues that resource exploitation is not necessarily self-limiting in that the carrying capacity of the environment can be increased and trade can provide for what is unavailable locally in terms of key inputs. Katzman also points out that economic behavior cannot be explained adequately by theories of energy value. Moreover, we would argue, changing world demand has continually redefined the natural endowments of Amazonia according to their economic value and brought new products to market. Although environmental degradation, such as deforestation, overfishing, and overhunting accelerated rapidly in certain moments, the set of natural (economic) resources demanded from the region appears to have expanded at a rapid rate also, probably even faster than extinction due specifically to product extraction. Whereas the set of subsistence resources involved (e.g., food, fibers, and construction materials) may be somewhat more restricted (i.e., defined by cultural norms), the number of natural products of the rain forest with potential commercial value appears to be very large. Furthermore, not all resources have been harvested in a non-renewable manner, a case in point being hevea rubber, the boom era's principal source of rubber. Finally, the thesis

---

[29] Bunker (1984) developed an argument that applies more specifically to the rubber era and why development was not sustained. The argument there centers on two ideas: the importance of labor scarcity as the main constraint on the transformation from extraction to plantation production of rubber and the insufficient local accumulation that resulted from the social relations of extraction. Earlier in Part I we also use labor scarcity as part of the argument for the plantation question, but we offer an alternate interpretation of local accumulation possibilities.

regarding the political ecology of underdevelopment relies, in the end, upon a purported failure in local surplus retention and on a combination of dependency arguments about unequal exchange and a lack of surplus generation because of the apparent inefficiency of extraction. The political ecology argument is perhaps more instructive in studying long-term trends in resource use in Amazonia than it is in analyzing the specific case of wild rubber and the Rubber Boom.

The Marxist Thesis of Blocked Development

Barbara Weinstein's (1983a, 1983b, 1985, 1986) work on the Amazon Rubber Boom has expressed in Marxist terms a coherent, empirically specific, and compelling view of why development was blocked in the region. She has argued that sustained economic development was frustrated not by surplus drainage from Amazonia but rather by the persistence of pre-capitalist relations of production. Specifically, an alliance emerged during the boom between rubber tappers (who owned the extracted rubber and controlled the means of production), and rubber traders (who controlled the exchange of rubber) that prevented the penetration of capital and proletarianization of labor. Pre-capitalist relations effectively blocked regional development by stifling capital accumulation, modernization of the wild rubber industry via plantation development, and thus the emergence of significant internal markets and other sectors (Weinstein, 1983a:135–35; 1983b:96, 263–65). A related (neo)Marxist view (Pennano, 1988) acknowledges the importance of untransformed relations of production but adds the dependistas' emphasis on the disarticulation of domestic underdeveloped economies caused by their international trade links.

Weinstein's works, in fact, contributed much needed attention to the actual conditions of rubber extraction and trade in Amazonia during the boom, and perhaps for this reason her explanation for the unsustained nature of growth and development during the boom era continues to hold considerable currency among Amazonianists (e.g., Schmink and Wood, 1992:45; Hecht and Cockburn, 1989:62). Unfortunately, her argument is deficient in several respects. In its barest form, the argument is tautologous: (capitalist) development did not occur because pre-capitalist relations were not transformed. But the crucial question then remains, why did pre-capitalist relations persist and effectively block development? Weinstein (1983a,b,1986) attributes the persistence to tapper resistance, rooted in an overriding need for personal autonomy and a natural disdain for wage labor, and to trader resistance, based on their determination to control trade. In her view, individual tapper preferences and choices produced *de facto* class resistance, which monopsonistic traders readily reinforced.

The argument we developed in Part I shows that Weinstein's arguments are problematic in several ways. What Weinstein sees as autonomy-seeking behavior of tappers cannot be distinguished from the autonomy inherent in the dispersed nature of wild rubber extraction, which made monitoring of labor effort prohibitively expensive. The argument for tappers' preference for autonomy over more remunerative arrangements (in wage labor plantations) thus can be sustained only by assertion. Moreover, the ability of traders to widely monopolize exchange is dubious given the low barriers to entry in trading, the mobility of river traders, and the abundant opportunities for pirate buying of rubber from tappers along the rivers. A more compelling explanation of the durability of the trader-tapper debt-merchandise contract is its relative efficiency when compared to other contractual arrangements in this decentralized, labor-scarce environment where transaction costs are high.

On empirical grounds, considerable evidence shows that surplus was indeed accumulated locally by private parties and by the state, and was available for investment in development-related activities in rubber or other sectors. The Marxist thesis of thwarted modernization is not sufficient to explain how such capital was accumulated or why its investment did not lead to sustained economic development in the region. In this sense, it joins the other two leading perspectives on the region's economic development experience in failing to focus attention on the actual patterns of capital accumulation during the dynamic and transformative boom period and how investment of such surplus shaped the evolving structure of the rapidly growing Amazonian economy.

An Alternate Theory—A Macroeconomic Model of Distorted Development

Corden and Neary's (1982) Dutch Disease model offers a coherent and compelling explanation for structural change in a boom economy by specifying the ways in which opportunities created by the boom sector influence resource allocation across other sectors of the economy. Specifically, the model helps us explain in the case of wild rubber how the boom became:

- the bane of other tradeable sectors (i.e., goods that can be imported and exported, such as most of those produced in agriculture and industry); and,
- the boon of the non-tradeables sector (i.e., goods and services that can only be produced locally such as real estate, construction, and many types of services).

As the boom played out over decades, the non-tradeables sector grew impressively while the tradeable sectors beyond rubber withered, creating a distorted regional economic structure. Prosperity in the non-tradeables sector came to rest on the single pillar of the rubber trade and, for reasons explored below, other tradeable sectors could not fill the economic void that would be created by the collapse of the Rubber Boom.

THE DUTCH DISEASE MODEL

At the heart of the Dutch Disease model is the boom sector, in which soaring prices of a commodity, a major discovery of a low-cost source of a resource, or a technological breakthrough that is not easily imitated by competitors have become a source of substantial economic rents. The high returns to this sector will be reflected in rising wages and/or capital earnings, which leads logically to a movement of productive resources into the booming sector. If no labor and capital flow into the region of the booming economy, then the movement of productive resources into the booming sector must result in a reduction of production of other tradeable and non-tradeable goods. However, if resources do flow into the region, at some cost, then this intersectoral movement of resources will be curtailed to an extent, though not completely, as the high returns of the boom activity compete away scarce labor and capital. The resulting movement of resources into the boom sector is known as the "resource allocation effect."

The shift of resources into the boom sector proceeds in a highly preferential way and thus has a very uneven impact on the other sectors. In particular, the non-tradeables sector is able to retain key productive factors by paying higher wages or generating higher returns to capital, because the local prices on these goods can be raised to offset these higher costs. The very "non-tradeable" nature of such goods means that the production, distribution, and pricing of these goods and services is local, and hence not be disciplined by competition from producers in other areas.[30] In contrast, the non-booming tradeables sector is not able to raise prices to retain labor or capital, because these goods and services can, in fact, be bought from other regions via trade. Thus, the ability to pay higher wages and capital costs allows the non-tradeables sector to attract local factors into employment and thus maintain production levels more readily than is possible in the non-booming tradeables sector.

The booming sector also generates "spending effects." Rising wages and returns in the boom economy push both incomes and thus expenditures higher, for both tradeable and non-tradeable goods. Spending effects are particularly apparent in the case of export booms, as in the case of Amazonian wild rubber, where the boom sector product is entirely exported. For similar reasons to the resource allocation effect, the effects of increased spending work differently in the two sectors. In the non-booming tradeable goods sector, price rises are again constrained by competition from imports. Higher expenditures on tradeables thus translates into more imports but probably little additional production of tradeable goods.[31] Meanwhile, in the case of non-tradeables, prices can rise with expanding demand, thus

---

30 On the margin, what is tradeable and non-tradeable actually may change as prices of non-tradeables in a given area rise.

31 For example, besides wild rubber during the Rubber Boom, tradeable goods production accounted for less than ten percent of total economic output in the peak decades of the boom, 1890–1910.

encouraging resources and investment to flow into that sector. In this way, the spending effect pulls even more productive resources out of the non-booming tradeable goods sector and into the nontradeable sector.

Resource allocation and spending effects described by the Dutch Disease model combine together to effectively direct investment and labor into the booming sector and the non-tradeable sector at the expense of the non-booming tradeable sector. By this process over time, an economic structure emerges that becomes highly reliant for its vitality on the fortunes of the boom sector, which becomes the principal source of revenue available to sustain the burgeoning non-tradeables sector. As long as the boom persists (or the returns to that sector persist), incomes rise and development opportunities are present. Once the boom fails, then changes in the economic structure caused by the boom may or may not begin to bind. If resources flow smoothly, or without adjustment costs, from the booming sector and the expanded non-tradeables sector to the non-booming tradeables sector, then even a temporary boom would be relatively unproblematic. When prices are high in the boom sector, resources will flow into producing that good and non-tradeables, and when prices of the boom good fall again, resources would flow back toward production of other tradeables, making "boom sector reliance" rather unproblematic. A boom under such circumstances may give rise to a one-shot increase in consumption, and then a return to pre-boom levels.[32]

A local developing economy becomes vulnerable to the Dutch Disease effects, however, when resources cannot be readily reallocated back to non-boom sector tradeables production, or when the loss of experience in the non-boom sector means foregoing important external economies that might have increased future returns in that sector. This latter effect—the problem of foregone external economies—though interesting in some contexts, is less relevant to our case of wild rubber where economic production was generally extensive and entailed rather simple technology. The former effect—the problem of adjustment costs—can be substantial where investments in the boom sector (or the non-tradeables sector) are irreversible or largely sunk in their nature. By "sunk," we simply mean investments where the capital or labor used are costly to either transform into other uses or to move to other locations for similar uses. A major decline in the price of the boom commodity signals economic agents to move resources toward production in

---

[32] In the case of smooth resource flows, inefficient intertemporal allocation of consumption can occur between the boom and non-boom eras (Woo *et al.*, 1994). In other words, the economy might undergo a "consumption binge" during the temporary boom, if either private agents had viewed the boom as a permanent one, and thus increased their consumption accordingly, or if the government obtained a sizable share of the boom surplus via taxation, and was unable for political economy reasons of multiple pressure groups to resist spending too much of the proceeds on consumption. Overall, given the temporary nature of the boom, economic welfare for the boom region would have been higher if future consumption possibilities had been generated by more savings and investment. Such effects, however, are minor in comparison to those associated with Dutch Disease.

other tradeable goods, but such a move may entail high costs. Such adjustment costs can render the resource less valuable than prior to the move. The potential for such losses associated with sunk costs can make the Dutch Disease effects especially problematic.

Sunk costs are relatively common and pervasive in extractive industries and the surrounding economy. Investments in extractive activities generally are made in remote locations where specific natural resource are abundant. Much of the equipment and infrastructure put in place for the extraction, storage, processing, and transport of the good may have little value outside of the particular activity, due either to geographic isolation or the specific features of the technology. In either case, the costs of moving the equipment and infrastructure to other uses is increased by the remoteness of the location and the limited value of the technology to alternate uses.

Also, the engineering demands of moving large volumes of raw resources (e.g., in mining and hydropower), require major investments to construct and maintain secure installations. Those set-up costs will be inherently sunk, because such facilities, in turn, would be costly to dissemble and move, probably more so than starting afresh with new equipment. In addition, raw material extraction and processing (e.g., separation, purification, etc.) often is most economic with large volumes of materials at relatively large scale. For this reason, investments for these activities will also tend to be rather "lumpy," which increases the absolute level of attendant sunk costs. Such "lumpiness" carries over into transportation—major transport infrastructure is needed to move and store the resource. The remoteness of the site of extraction and the unique physical characteristics of the resource can strongly limit the potential for such infrastructure to be used for alternate activities.

Sunk costs in resource extraction industries reach beyond capital investments in facilities and transport. Where an industry must recruit, relocate and provision a labor force, and establish ancillary services, in a remote area, sunk costs are also entailed. Investments in ancillary facilities and services that are provided for labor, and directly related only to the boom industry, would imply further sunk costs. Although when the extractive boom fails, labor may bear the brunt of the sunkness of such investments, the costs themselves are associated with the extractive industry.

In summary, extractive industries are prone to high levels of sunk costs, and the economic consequences can be rather severe when the resource boom goes bust. Assets suddenly lose a good portion of their underlying value because of their sunkness, and this contraction dries up both the prospects for reallocation of capital to alternate tradeables production and, perhaps just as important, the consumption power necessary to sustain the value of assets in the non-tradeables sector. Even worse, many of the investments in non-tradeables are themselves likely to be fairly sunk: houses, commercial buildings, restaurants, theaters, hotels, bars, etc. have asset values that are tied to the fortunes of the region, and such values will decline dramatically in the wake of the boom if other alternate economic activities cannot be made profitable. In this respect, the economic structure shaped by an extended resource boom—with its vibrant non-tradeables sector resting on the fortunes of the booming resource sector—can prove to be extremely fragile if the boom fails. Much of the asset values built up in the boom will essentially "evaporate" because of the sunk cost nature of the investments in the boom and non-tradeables sectors.

THE DUTCH DISEASE MODEL APPLIED TO THE AMAZONIAN RUBBER ECONOMY

The Dutch Disease model was conceived originally with developed economies in mind (e.g., Netherlands), ones with a diversified tradeables sector that already included active agricultural and manufacturing sub-sectors which might be disadvantaged in a permanent way by a resource boom. The economy experiencing the boom is often assumed to have either no or relatively inelastic inflows of productive factors, which means that higher returns to local factors pulls labor and capital into the boom sector and, in some cases, into the non-tradeables sector, and sharply away from the non-booming tradeables sectors. Conditions in Amazonia were clearly different. Prior to the boom, the tradeables sector was constituted almost entirely by agriculture, based on some export-based plantation crops and some subsistence crops. The regional economy was small in economic terms and essentially confined geographically to the Belém area at the mouth of the Amazon River. Labor and capital were both extremely scarce in the basin, relative to the demands that would be introduced on these factors by the Rubber Boom and the concomitant rapid expansion of overall economic activity. Thus, labor and capital flowed rapidly into the region, drawn by the attractive returns promised to successful participants in the boom.

These differences between conditions in Amazonia and those in more developed regions mean that the concerns about structural distortions in the economy raised by the Rubber Boom are rather distinct. The classic concerns, for example, about de-industrialization with a resource boom are not relevant to Amazonia during this historical period, for prior to the boom, the region as a whole had very limited industry in place. Although labor and capital were indeed pulled out of other productive sectors, such as agriculture, it was the massive inflows of new capital and labor that proved to be most influential in shaping the regional economy. Rather than inducing structural change in a well-diversified and integrated economy, high returns to rubber and the inflows of new resources essentially gave rise to the formation of a boom economy, with its expansive resource sector and a burgeoning non-tradeables sector. These conditions raise two important empirical issues for our analysis in the next two chapters. First, to what extent did the resulting economic dynamics and structure in Amazonia reflect the logic of the Dutch Disease model, adapted for this context? And second, what role did sunk costs play in thwarting potential adjustment strategies when the boom failed? These questions can be answered through careful study of the investment patterns and economic structure of the boom era, and of the barriers to economic adjustment that may have limited the options for a smooth transition when the Rubber Boom was over.

Evidence on Boom Investment and the Role of the State During the Boom

Before we turn to our study of investment and the role of the state during the Rubber Boom, we must first briefly discuss an important constraint on our efforts. Our search through the literature for relevant works to inform a closer look at these issues was rather disappointing. This paucity of information is due, in part, to the previous emphasis in the literature on the social relations of extraction and trade within the rubber sector. Fortunately, recent works by Roberto Santos and Barbara Weinstein do offer some empirical information on the relationship between the rubber sector and the rest of the economy, which are quite useful. Santos (1980) pieces together evidence on the structural evolution of the regional economy during the Rubber Boom and beyond. Weinstein (1983b) stresses the tensions that arose between the ascending rubber sector and the region's traditional agricultural interests, drawing on several informative archival sources including commercial and individual notarial records.

Descriptions of Amazonian urban development, like Bradford Burns' (1965) account of the rise of Manaus, provide useful secondary material for making inferences about investment patterns. In terms of the state, snapshots of the taxation and expenditure patterns of regional and federal governments can be found in LeCointe (1922), Fuentes (1908), Pearson (1911), and others. Nonetheless, the scarcity of published empirical information does limit our analysis of private and public investment patterns and the evolution of the structure and performance of the overall Amazonian economy. We must rely therefore in the next two chapters more on inferences and deductive reasoning than might be expected. We do so with the hope that new avenues will be opened up for fruitful archival research on these issues.

Chapter 8

PATTERNS OF PRIVATE INVESTMENT DURING THE RUBBER BOOM

Introduction

Our discussion of investment patterns during the Amazon Rubber Boom begins with a description of the actual types of investments made by the private sector in the boom sector and the non-booming tradeables sector. A closer look at the basic features of these investments reveals the limited utility for alternate economic activities inherent in boom sector investments, the few valuable linkages they encouraged, the minimal investment activity in non-boom sectors, and thus the fragile foundation provided by these investments for sustained Amazonian development. Our subsequent discussion of the Dutch Disease effects demonstrates how imbalanced the regional economy became, with wild rubber exports and non-tradeable goods and services accounting for almost the entire regional output at the peak of the boom. Given the substantial sunk cost that accompanied most of the investments in these two sectors, it becomes readily apparent why successful economic adjustment in the bust era proved to be so difficult.

Investment in Rubber Extraction and Transport

Most private investment during the Rubber Boom went into the operation and expansion of wild rubber collection and trade. As the boom developed, participants invested in new estates further upriver, in setting up tappers to work the estates with supplies, in transporting and transshipping goods along the rivers, and finally in collecting, storing, grading, and trading the rubber at major ports. Although no reliable estimates of sectoral investment behavior have been constructed in the literature, a general sense of the level of expenditures and the relative sunkenness of such investments can be acquired from studies from the era.

To establish and run a rubber estate, considerable investment was required in claiming the property, laying out the rubber trails, and in recruiting and provisioning tappers. The main physical investment on an estate involved laying out the tapper's rubber trails (estradas). At a cost of $100–200 per estrada, this investment was highly product-specific and essentially "sunk," in that estradas had no other economic use. Such costs, however, were relatively small compared to the working capital to bring tappers to the estate and to set them up to work. Recruiting, transporting, equipping and supplying tappers for their first season required about $500, plus $200–300 per year thereafter for provisions, all of which were provided as loans. At $500/tapper, we estimate conservatively that such investment in rubber estate activity in the Brazilian Amazon, where about 150,000 tappers worked (per Santos, 1980:66), would represent about $75 million in working capital alone. At any given time, the capital value of such loans made to tappers depended on the price of rubber to sustain future repayment because no other extractive activity in the Amazon, particularly in the upper reaches, could generate sufficient earnings to support the costs of these provisions to tappers. Thus, the value of these working capital investments in labor, though neither inherently product-specific nor sunk like estradas, rested on the vitality of the rubber sector.

Rubber transport required two types of investments—vessels and infrastructure. The main forms of transport were steamers, launches, small boats and canoes, whereas passages for portage, warehouses and other trade-related buildings, and ports made up most of the infrastructure. Boats—the classic example of mobile capital—could be used to service other industries or move to other regions. But mobility of such capital could cut either way: allowing support for alternate economic activities, or exit from the region. The fact that many of the boats would have been too costly to relocate to other areas of the world (indeed, some were very expensive to situate in the upper rivers in the first place) meant that after rubber went bust, a general abundance of shipping capacity existed in the region. How well the boats were serviced and maintained, however, was another matter; many were abandoned or lost to the river by the late 1910s and early 1920s.

Investments in infrastructure (wharves, warehouses and portages) were sunk, highly dispersed but not product-specific, except in terms of location. Thus, like loans to tappers, the salvage value of such investments was tied directly to the trade in rubber. When the boom collapsed, no subsequent extractive activity in the Amazona could generate a similar volume of export-import trade and much of the infrastructural capacity became superfluous. Moreover, infrastructure of all types deteriorated rapidly in the humid tropical environment, especially outside of the major port-cities. Portages and estradas became overgrown. Floods washed away ports. Warehouses and buildings fell into disrepair. Trading posts closed, and traders moved on.

In the main port-cities of Belém, Manaus, and Iquitos, rubber was not refined beyond grading and sorting by quality. Urban investments in physical capital directly related to rubber were mostly for transshipment facilities like ports, trading houses, grading facilities, and markets. Port facilities, the most costly of these investments, were publicly subsidized to a significant degree, although often privately operated. Urban infrastructural investment, too, depended on a high volume of trade to justify its maintenance, and much of the cities' infrastructure was also neglected after the bust.

Most of the direct capital investment in rubber extraction and transport was spread out spatially yet concentrated in the form of working capital, information, and transportation networks that were highly specific to the rubber sector. Most physical capital was sunk (because of the high costs of moving it elsewhere), and much capital was product specific, either inherently so or because so few alternative extractive goods from the forest could sustain even the upkeep costs for more flexible investments like boats and infrastructure that had been used by the Rubber Boom. Thus, when rubber prices collapsed, much of the physical capital associated with the rubber industry lost its value and fell into disuse. Adjustment options for investors were strongly limited by the types of investments they had made in the region, by the associated high costs of transforming or transporting them, and by the limited returns to other highly dispersed extractive activities. Faced with inevitable losses, many managers and traders left in the wake of the boom, further reducing the available expertise for promoting alternate economic activities in the Amazon.

Production Linkages

Wild rubber extraction generated few local linkages. Tapping required minimal fixed physical capital. Establishing estradas largely involved a labor expense to find the wild rubber trees and lay out the best circuit for tapping. Implements were limited to a few manufactured items, such as tapping hatchets, cups for latex collection, pails, and guns for hunting and self-protection. In general, these items were cheaper to import than to produce in the region. The limited scale of demand, a highly decentralized population, and the high input cost environment of the Amazon worked together to disadvantage prospective local producers of tradeable goods. A major exception on the input side was the manufacturing of soap, which was used to coagulate certain types of rubber latex (such as caucho). As early as 1862, twenty four soap manufacturing establishments existed in Belém (Santos, 1980:188). One other linked industry that arose in the region was the construction of wooden crates for rubber shipment. In the transportation sector, the major linked industry was the provision of wood for steamers, which offered additional labor opportunities but little investment or diversification for the post-boom era.

The other manufactured goods that served the needs of rubber trade participants were mostly common consumer items, such as cashew wine, rum, chocolate, rice, beef, and furniture. Production of such consumer goods took advantage, like soap and crate production, of local raw materials and relatively high unit transportation costs. The primary market for consumer goods was the population in urban centers which had grown up along the Amazon River and its tributaries at the base of the rubber trade.

On the downstream side of the industry, rubber manufacturing was undergoing rapid and complex technological changes in the industrial centers of the North America and Europe.[33] Local manufacturing ventures in Amazonia were distant from innovations and information on changes in consumer needs and industrial user specifications. The rapid pace of technological innovations in industrial uses of rubber effectively left behind the incipient and artesanal rubber processing industry that had developed in the state of Pará as the boom began.[34] Only after the boom, when some production technologies had become more standardized, did rubber manufacturing return to the region.

Because the technology of rubber extraction was both simple and specific to rubber, no significant technological or learning spilled over to other extractive sectors. Probably the most important lateral linkages generated by the sector were the extensive transportation networks established throughout the basin and the development of major urban centers. On the whole, though, the sector-specific nature of rubber industry investments left the economy vulnerable to a bust in rubber prices. Investment in linked activities did not generate diversification into more independent economic sectors, and the capital invested in rubber extraction and transport could not be transferred easily to other economic activities or sectors.

Alternatives for Private Investment in a Boom Sector Economy

Investment alternatives for the private sector during the Rubber Boom were biased against other tradeable goods and toward non-tradeable activities, particularly urban real estate and service industries. Rural investment in tradeables, unrelated to rubber extraction and transport, was limited and occurred mostly on smaller or downriver estates where owners sought to delineate their property rights, often by clearing land and planting perennial crops. At the same time, regional accumulation of wealth was nevertheless spectacular during the boom. Belém, Manaus, and Iquitos became major urban centers, boasting the most modern of amenities, services, and consumption activities. But the future value of this wealth continued to

---

[33] The pace of invention and innovation around rubber was remarkable. One has only to browse through the advertisements, patent reports, and news releases published in trade journals like *India Rubber World and Electrical Trades Review* (1889–1899) and its successor, *India Rubber World*, to appreciate what novel and diverse possibilities were catalyzed by rubber for both industry and consumers in the late 19th and early 20th centuries.

[34] See Edwards (1847:179–180) on the early rubber processing industry near Breves.

depend on the price of rubber, because the economy was so little diversified and so much of the accumulated urban wealth was held in real estate and property rather than in more potentially productive facilities or savings.

Reasons on both supply and demand sides explain why profits or savings from the rubber industry were not generally channeled into producing other tradeable goods. Geographic factors were paramount: specifically, the remoteness of the locale vis-a-vis markets in the North, and the associated high costs of inputs made tradeable goods production unattractive. Basic inputs had to be shipped long distances into a tropical environment, implying high shipping costs, especially given that many of the basic inputs were often many times the volume and weight of the finished or semi-finished good. High costs in non-rubber extraction and agriculture, partly due to high wages required to keep scarce labor out of the renumerative rubber sector, meant that even many of the raw inputs could not be procured locally at low cost.

On the demand side, low population density and a relatively small population meant a limited market for developing import-substitution manufacturing ventures. Such efforts were further discouraged by the potential competition from imports that could be backhauled on empty rubber boats at relatively low unit cost from the major industrial areas, where economies of scale or better production conditions reduced costs below what was possible in Amazonia. Hence, the tradeable goods sector faced formidable barriers to being competitive on several fronts related to the geographical isolation of the Amazon basin relative to world centers of demand, the type and availability of specific endowments of the region, and the market realities of the Rubber Boom economy.

Evidence of Dutch Disease Effects During the Rubber Boom

The Dutch Disease effects of the Rubber Boom can be seen clearly in Table 4. The table presents the total value in current *contos de réis* of total economic product and the relative contributions of the various sectors and sub-sectors in the regional economy of the Brazilian Amazon during the boom. In the primary sector, the effects of the boom on agriculture are quite apparent. Whereas the boom in rubber is reflected in the expansion of the second sub-sector ("plant extraction") from 37,914 contos in 1890 to 197,811 in 1910 and a 5 percent increase in the proportion of economic activity accounted for by extraction, the total product of agriculture is essentially at the same level in 1910 as in 1890. More to the point, agriculture's share of total product dropped from about 8 percent to 2 percent during this period.

Apparent stagnation in the agricultural sector actually obscures two other significant phenomena. The first is revealed by the 1900 figures which indicate that the product of agriculture doubled in the decade of the 1890s and then contracted by the same magnitude in the first decade of the 1900s. The expansion is overstated because of the inflation that occurred during the

1890s, but the contraction of the 1900s corresponds to the period of highest rubber prices, with the peak rubber prices of the era coming in 1908–1910. This contraction is consistent with the decline that would be expected in other tradeable sectors when the boom product's price is rising. Second, growth experiences varied across different commodities, with output of export crops like rice and cacao declining sharply while the production of some crops for local consumption (e.g., manioc, tobacco, and rum) rose to meet local demand during the boom. Although the agricultural sector did not experience wholesale collapse, as some observers of the era might believe, the Dutch Disease effects on agriculture are decidedly apparent in its relative decline.

In the secondary sector, industrial activities expanded over the two decades, but growth in aggregate terms was only one-tenth of that in the primary sector. At the peak of the boom in 1910, industry accounted for only 3 percent of total economic product in the Amazon region, and much of that was in industries that processed raw materials for either rubber extraction and transport, such as soap to coagulate latex and saw mills to construct shipping crates for rubber, or the urban non-tradeables, such as whitewash and wood for painting and construction (Table 4).

Although many of the larger industrial ventures collapsed with falling rubber prices after 1910, overall industrial activity actually continued to expand in the decade of the rubber bust. Such growth can be attributed both to the decline in labor and capital costs associated with lowered earnings in the wild rubber trade and to the fact that food processing firms, clothing manufacturers, furniture makers, and chemical producers found a market among the urban population for lower quality import substitution items because local incomes were no longer high enough to afford higher-quality manufactured imports. Indeed, a 1920 census of industrial concerns in the Brazilian Amazon found that more than half of the operations had been established during the 1910s, with the most rapid expansion occurring in food processing and clothing manufacturers, where 82 of the 156 new firms were found (Santos, 1980:189). Such figures are vulnerable, of course, to the potential criticism that, over time, the natural exit of firms would tend to bias the proportion of establishments accounted for by new firms. However, earlier data from the state of Pará on industrial establishments (see Santos, 1980:188) suggest the existence of only a few food processing and clothing manufacturers in 1892. As such, these data support our inference that the import-substitution sector expanded significantly in the years following the crash, over the view that the new firms were merely replacing old ones.

During the boom, the main growth area besides rubber was in the tertiary sector, especially in wholesale and retail trade and government

TABLE 4. TOTAL REGIONAL ECONOMIC PRODUCT BY SECTOR, BRAZILIAN AMAZON, 1890–1920

Economic sector	**1890**		**1900**		**1910**		**1920**	
	CONTOS DE REIS*	%	CONTOS DE REIS	%	CONTOS DE REIS	%	CONTOS DE REIS	%
Primary Sector	*53,953*	*50.8*	*181,040*	*48.8*	*218,287*	*44.9*	*123,507*	*35.8*
Agriculture	8,142	7.7	20,833	5.6	9,593	2.0	31,251	9.1
Plant Extraction	37,914	35.7	141,484	38.1	197,611	40.7	57,182	16.6
Animal Products	7,896	7.4	18,723	5.1	10,883	2.2	35,074	10.1
Secondary Sector	*548*	*0.5*	*6,222*	*1.7*	*19,605*	*4.0*	*24,632*	*7.1*
Mineral Extraction	0	—	—	—	—	—	59	0.0
Industry	147	0.1	3,054	0.8	15,684	3.2	20,579	5.9
Civil Construction	401	0.4	3,168	0.9	3,921	0.8	3,994	1.2
Tertiary Sector	*51,721*	*48.7*	*183,877*	*49.5*	*247,941*	*51.1*	*197,450*	*57.1*
Wholesale and Retail	36,003	33.9	102,216	27.5	149,606	30.8	134,595	39.0
Government	7,793	7.3	51,220	13.8	53,270	11.0	28,870	8.3
Other Services	7,925	7.5	30,441	8.2	45,065	9.3	33,985	9.8
Total	106,222	100	371,139	100	485,333	100	345,589	100

Source: Santos (1980:178) with the explanation of the product estimates presented in the methodological appendix.

*Contos de réis was the Brazilian currency in that era.

Photo 12. Palace of Justice, Manaus, ca. 1910

NOTE: The cost of transferring a case of rubber from this dock to the pier via the aerial cables was $0.38 or the equivalent of shipping the same case from New York to Australia (Pearson, 1911:100)

Photo 13. Floating Docks at the Port of Manaus with Aerial Transport Cables, ca. 1910

services.[35] When combined with rubber exports, trade and government accounted for over 80 percent of total economic product during the peak years of the boom, whereas industry and agriculture accounted for only about 5 percent (Table 4). In fact, if we include the other service sector as a non-tradeable, then rubber and non-tradeables, in total, accounted for more than 90 percent of the total domestic product in the Brazilian Amazon in 1910.

Bradford Burns' account, "Manaus, 1910: Portrait of a Boom Town," describes the urban achievement of one of the world's leading centers of wild rubber exports and serves to highlight the central role of urban non-tradeables in the boom economy (see Photos 12–13). Located 900 miles up the Amazon River, Manaus had been a relatively poor and undeveloped, jungle river port of 5,000 inhabitants in 1880. Only thirty years later, at the peak of the boom,

> An excellent system of waterworks, efficient garbage collection and disposal system, electricity, telephone services, handsome public buildings, and comfortable private residences attested to the modernity of the city .... This capital of 50,000 inhabitants was bound together by a steel band of fifteen miles of electrical railway, whose streetcars came and went from the praça [main plaza]. (Burns, 1965:401).

He also depicts the high life of Manaus, which included fine French fashion, fancy restaurants, sporting clubs, opera and music halls, movie theaters, high quality newspaper and periodical publications, and numerous and well-funded public and private schools, along with the seamier side of boom-town life, such as gambling halls, brothels, and bars. Similar types of urban development occurred in Belém and Iquitos, which exemplify the bias of a boom-sector economy toward investing in non-tradeables.

Overall then, huge imbalances were created in the regional economic structure of Amazonia with the success of the wild rubber industry. The tremendous expansion of the rubber and urban non-tradeables sectors during the Rubber Boom, as suggested above, need not have presented a major problem (at least in principle) for the economic development of Amazonia. If the price rise in rubber had proved to be permanent, or if capital, labor, and other productive resources could have shifted easily back into the production of competitive tradeable goods after rubber prices fell, then this expansion would have meant rising incomes and welfare for the Amazon. In the case of a permanent price rise in rubber, participants in Amazon rubber

---

[35] As a point of possible clarification here, the retail and wholesale trade is actually considered to be part of the non-tradeables sector. The value added to the economy by this sector is considered to be the effort undertaken locally by the participants involved in the pursuit of trade, and not the value of the imported goods themselves. Local services generally are not tradeable outside of the region where they are provided. A local merchant reselling imported goods, for example, or a river trader transporting the goods to patrons and tappers, could not be considered to be trading (or being able to trade) such services in the world market; instead, they are seen to be providing an essentially local service.

production would have continued to enjoy the gains from specialization and trade. In the case of costless adjustment of resources into other tradeable goods production, income gains and capital accumulated during the boom would have resulted in both increased consumption and improved future production possibilities after the boom.

The problematic nature of the Amazon Rubber Boom, however, lay in the high cost and difficulty of shifting productive resources out of the booming and non-tradeable sectors into other tradeable goods. Many of the capital investments made in rubber extraction were both product-specific and sunk in nature. No alternate high-value extractive activities were available to sustain the infrastructure developed to serve the highly dispersed rubber industry. For example, in many areas of Brazil, extractors turned to the collection of Brazil nuts (*Bertholletia excelsa*) but revenues were only a fraction of those from rubber during the boom (LeCointe, 1922:I:447–49, 461), and economic rents were even less.[36] The adjustment problem was perhaps even worse with respect to the non-tradeables sector. Moving major urban infrastructural investments, commercial real estate, even stocks of fine foods for resale in other markets would have faced formidable transport costs. It is no accident that even today one of the most telling legacies of the boom era in Amazonia are the decay-prone façades in Manaus and Iquitos on buildings that once were touted to rival the grandeur of Europe's finest.

When rubber prices collapsed after 1910, Amazonia's economic structure changed both rapidly and substantially. A major decline occurred in the overall gross domestic product: in the Brazilian Amazon, the regional economic product dropped by 30 percent. Rubber revenues fell from 197,811 to 57,182 contos de réis. By 1920, the share of industry, agriculture, and animal products in the region had risen to 25 percent while the share of rubber, trade, and government had contracted to 65 percent (Table 4). The expansion in industry and agriculture, which amounted to an increase of 70,000 contos de réis between 1910 and 1920, could only cover about one half of the decline in wild rubber revenues during that decade. The economic product of the tradeables sector for 1920 was 40 percent lower than in 1910.

Interestingly enough, although retail and wholesale trade, the second leading sub-sector behind wild rubber during the boom, also declined during the 1910s, revenues for this sub-sector fell only by about 10 percent. As a result, retail and wholesale trade actually represented a greater percentage of total economic product in 1920 than at any time during the Rubber Boom. The fact that this sector did not experience a more substantial collapse suggests that local purchasing power from earnings generated and saved during the boom helped to sustain the non-tradeables sector well beyond the

---

[36] For a discussion of Brazil nut industry in the 1910s and 20s, see LeCointe (1922:I:453–465) and Pierrot (1924), respectively. De Mello Petey (1972) provides a more general discussion of economic conditions, including Brazil nut extraction as well as agricultural activities, in the Brazilian Amazon during the post-boom period.

crash. This, in turn, also supports our earlier argument for substantial local surplus retention during the boom.

In sum, when rubber prices collapsed in the 1910s, the underlying fragility of the boom economy was exposed. Investments in rubber extraction and infrastructure rapidly lost value, in part because they could not be transferred readily to other tradeable activities. At the same time, activities in the non-tradeables sector quickly lost the large inflows of income from the rubber trade that had sustained them. Although agriculture and industry began to expand, both started from a low level of technology and a minimal productive base after years of limited attention.

The combination of the collapsing boom sector and a declining non-tradeables sector led to massive depreciation of capital, as values of non-transferable investments from the boom era plummeted. Years of income growth, private investment, and apparent economic development had been lost. A fragile boom economy of rubber and an overdeveloped non-tradeables sector gave way to an economy based on activities yielding much lower returns—agriculture, forest product extraction, and local industry. Even today, real per capita incomes in parts of the basin have yet to recover to levels during the boom.

Alternatives to Boom and Bust

Our analysis of the Rubber Boom economy and the Dutch Disease impacts of the boom inevitably leads us to the question of what alternate development strategies might have been pursued during the boom. What strategies could have served to cushion the fall, or to have left Amazonia in a better position to pursue more sustained growth and development? Could other tradeable sectors have been nurtured during the boom? Or, could funds have been set aside for a rainy day to insure against a major decline in rubber prices? Perhaps taxes on non-tradeables could have encouraged more private savings and investment instead of the high consumption evident in Burns' description of Manaus in the boom? In approaching these and related questions, it is important to recognize just how long the boom dominated the regional economy.

The Amazon Rubber Boom was one of the longest natural resource booms in recent history. Depending on the date chosen for mark its beginning, the boom lasted somewhere between thirty and fifty years. We join Weinstein (1983b) in considering 1860–1910 as a fair representation of the length of the boom, with the first 10 years or so being viewed as rather incipient. As such, the Rubber Boom was, for example, about three to four times the length of the ten year boom in oil prices and foreign lending that occurred between the mid-1970s and mid-1980s—a boom that generated numerous episodes of Dutch Disease in countries around the world. Shorter commodity price booms were also known in this period.

The long duration of the Amazon Rubber Boom and the fact that the boom served to integrate much of this vast region into the world economy via labor, capital, and trade inflows together made that the challenge of developing alternate strategies both difficult and fundamentally distinct from those in countries which have more recently experienced Dutch Disease effects. The regional economy, for the majority of people living in Amazonia, was created by the Rubber Boom. They had little, if any, experience, in the region under a fundamentally different price-return environment—and if the boom environment had been altered substantially, especially in the early years of the boom, many of them probably would not have come in the first place. Thus, alternative policies for confronting the economic fragility created by the boom's incentives for investment are probably only worth discussing in its later years, after considerable in-migration and economic expansion had occurred in the region.

As the boom advanced into the 20th century, the challenge of promoting alternate development strategies would have become particularly formidable. By the turn of the century, the boom had lasted long enough for two to three generations of residents to have experienced boom conditions for a good portion of their lives. After such a long period with wild rubber as the prime sector in the economy, people were ill-prepared to imagine the consequences of an unprecedented crash in rubber prices. The sense that wild rubber was both the past and future of Amazonia would have been hard to contest, especially as rubber prices accelerated in the early 1900s. With much money to be made in wild rubber, and with so much growth to be realized in the regional economy, few residents would have been inclined to pessimism. More important, it is hard to imagine what interests or classes in the private economy would have come to support the remediation efforts of a private or public sector naysayer.

Perhaps the only actor during the Rubber Boom that came to possess the financial resources and power to markedly influence incentives for private investment was the state. In the next chapter, we turn our attention to the role of the state in the regional economy. Of particular interest are the types of tax and expenditure patterns pursued by regional and national governments as well as to the types of incentives for investment that state initiatives helped to foster. Our analysis suggests the ways and extent to which the incentives created by the boom economy, under the influence of the Dutch Disease, were moderated or reinforced by the state.

Chapter 9

THE STATE AND RUBBER RICHES

Introduction

The wild rubber industry during the boom brought unparalleled prosperity not only to the private sector in Amazonia but also the regional governments that shared the basin. In the previous chapter, we suggested that, at least in principle, the state may have been in the position to employ such new found wealth to manage the structure of the regional economy for more sustained and balanced growth. The role of the state therefore becomes crucial to understand.

Prior accounts of the Rubber Boom offer us only limited insight into the nature and activities of the state during this crucial period; emphasis instead is placed heavily on the dynamic private sector. To many observers, the state was an opportunistic, complacent and, at times, inept player which left the rubber industry to do its business, satisfied by the absorption of rubber revenues though stirred to action periodically by the need to secure territorial claims (Resor, 1977:35; Walker, 1987:72; d'Ans, 1982:180).[37] The peril of plantation rubber was apparently recognized too late and, despite efforts to the contrary, the state could affect only limited change in the industry as rubber prices declined precipitously under competition with Asian plantation rubber. Despite this general tendancy to overlook the role of the state, one scholar (Weinstein, 1983b) does shed helpful light on the politics of wild rubber in the Brazilian Amazon during the era.

In this chapter, we argue that the rise of the lucrative rubber trade marked a period of intense development and transformation of relations between the state and Amazonia. The state aggressively sought and captured substantial rubber revenues that were thus available (at least potentially) for promoting regional development through public investment in industrial-

---

37 Resor (1977:35) writes, "As long as the rubber gatherers paid their taxes, the governments of Brazil, Peru, and Bolivia seemed willing to sit by and watch their treasuries fill up."

ization, diversification, and economic reform. We suggest, however, that much of the state's rubber revenues, rather than being destined for reform or diversification, was absorbed in prosecuting and protecting territorial claims, establishing regional administration, and developing local infrastructure and services to support the rubber trade. Such expenditures reflected strong emergent political and economic pressures on the state to direct rubber revenues back into the booming sector of this relatively free, open and emergent economy. These pressures reinforced patterns of private investment and heightened the regional economy's vulnerability to a collapse in rubber prices. The centrality of public taxation, expenditure, and policy in understanding the development path and legacies of the Amazon Rubber Boom underscores the need for more attention to the role of the state during the era.

State Revenues and the Capture of Rents

Amazonia was endowed with a natural monopoly on the world's finest wild rubber that endured for half a century until large quantities of cheaper Asian plantation rubber came to dominate world markets. States in the Amazon basin took advantage of this favorable market position to capture substantial rents from the trade through taxation at custom houses in the developing port-cities and towns along the Amazon River and its tributaries. Duties were collected on exports of wild rubber and other extractive products as well as on goods imported to the region. Such duties were very effective given the isolation of Amazonia, the limited development of local markets for food and other supplies, and the foreign nature of demand for wild rubber. The direct cost of collecting duties was probably only a small fraction, certainly less than 10 percent, of the revenues generated by the customs houses.[38]

Duties levied on Amazonian exports and imports contributed the bulk of government revenues captured locally by states during the boom, often more than 90 percent. Frequently, import duties were substantially higher than duties on exports—up to five times the duty on exported rubber—and although the value of exports typically exceeded imports (primarily consumer goods) during the boom, the largest share of government revenues from the rubber industry came from duties on imports rather than on exports. Typically, duty rates rose with rubber prices, which varied among the Amazonian countries according to relative market power, and were maintained well after the boom's peak by Brazil, the region's main producer. At the peak, merchants paid 100 percent import duties on goods brought to

38 According to the proposed 1906 budget for the Department of Loreto (which contained the principal customs house for Peruvian rubber exports at Iquitos), the customs service was to be granted about 6.4% of the revenue expected that year (Fuentes, 1908:I:271–287).

the Brazilian Amazon, and all such proceeds were destined for the coffers of the federal government. Import duties in neighboring countries of the Upper Amazon were considerably lower: merchandise imported into Peru was taxed at rates of 8–30 percent *ad valorem*, depending on the type of good. Imports to Bolivia were duty free, and Brazilian tariffs were adjusted along the western borders in an attempt to discourage a vibrant transboundary contraband trade that developed during the boom.

Although export duties were lower than rates on imported goods, the duty on wild rubber was substantial, particularly when compared with government taxation of other natural resources elsewhere.[39] Export duties on Pará fine rubber at the peak of the boom varied significantly across the basin, from the state of Pará (22 percent *ad valorem*) and the state of Amazonas (19 percent), to Bolivia (14 percent), and Peru (14 percent) (Pearson, 1911:58,98,142,160). In Brazil and Peru, revenues from export duties were destined for regional rather than national coffers, and municipalities in Brazil also levied a 1 to 2 percent tax on exported rubber (Pearson, 1911:58–59).[40] In the federal territory of Acre, the Brazilian national government imposed a 15 percent export duty on rubber (Pearson, 1911:165). Such high export and import duties provoked persistent complaints and conflicts both within and among countries in the basin.

Total revenues captured by Amazonian countries during the Rubber Boom were probably very large in today's terms, perhaps even "gigantic" as Randolph Resor (1977:350) has suggested. Between 1902 and 1910, federal and state governments in Brazil received in duties approximately 18.5 percent of the total value of imports and exports from the region, about $25 million/year.[41] Municipal duties and taxes would have added approximately 10 percent (i.e., $2.5 million/yr) to government revenues destined to federal and state governments (see Santos, 1980:193). In Peru, the state captured over the same period 10.6 percent of the total value of rubber trade, equivalent to some $7.02 million, two-thirds of which came from duties on imports.[42] Duties collected on rubber exported from Bolivia for the period totaled $2.74 million.[43] A conservative estimate of the surplus captured in Brazil, Peru, and Bolivia by all levels of government in 1910 at the peak of the Boom would be $47.4 million.[44] Over the period from 1880–1920, as

---

[39] For example, around 1900 and for a few decades into the twentieth century, *ad valorem* rates of government taxation for bauxite and banana exports elsewhere in Latin America were about 4% and 1%, respectively.

[40] Derived from LeCointe (1922:II:405, 424, 427). See LeCointe's (1922:II:432–434) discussion of frivolous municipal taxation.

[41] Derived from LeCointe (1922:II:405, 424, 427). U.S. dollars converted from local (or other foreign) currency at the official rate of exchange for that year, according to U.S. Mint (1890–1920).

[42] Derived from Maúrtua (1911:27, 28); Pennano (1988:204).

[43] Derived from Ballivián and Pinilla (1912:248–249).

[44] We estimate the surplus captured in 1910 from the wild rubber trade in Brazil as $45,346,290 (see Santos, 1980:193 using $0.33 per milréis), in Peru as $1,341,207 (see

much as $500 million in state revenues (the equivalent today of at least $5 billion) may have been generated by the rubber trade in the Amazon basin.

Although secondary information on the disposition of state revenues throughout the basin is limited, some data are available on the general balance of revenues and expenditures. In Brazil, the distribution of duty revenues favored the central government, which captured perhaps half of total revenues, followed by the state governments of Pará and Amazonas securing about one-third of revenues, and their municipalities receiving less than one-fifth (Santos, 1980:193; LeCointe, 1922:II:412,440). Federal expenditures in the Brazilian Amazon were relatively modest, on the order of 10–25 percent of received revenues, whereas state and municipal governments sought to match expenditures to revenues.[45] In Peru, the Department of Loreto received the revenues collected by the customs house at Iquitos and proceeds appear to have been used primarily in the region with any surplus going to the federal treasury (Fuentes, 1908:I:290).

Public Expenditures

With substantial revenues available for disbursement and growing demands by regionally-based groups for sharing the growing public wealth, the state became a major investor in Amazonia with the potential to shape the expanding regional economy and guide development through public spending and investment. In the Brazilian Amazon, for example, public expenditures by all levels of government climbed from the equivalent of $6.11 million in 1890 to $12.4 million in 1900 and $25 million at the peak of the boom in 1910.[46] State and municipal governments consistently budgeted for expenditure of their entire projected annual revenues. Although few comprehensive and detailed accounts of state spending during the boom are available, we surmise from the secondary literature that much of the rubber revenues was destined for serving geopolitical ends, establishing the state apparatus in the region, and promoting the rubber trade. As revenues rose during the boom, public spending increased and centered ever more tightly on the rubber sector.

Bonilla, 1976:226 using $4.8665 per pound sterling), and in Bolivia as $747,062 (see Ballivián and Pinilla, 1912:248–249 using $0.389 per boliviano).

45 See LeCointe (1922:II: 412,440); Santos (1980:194); Ministerio da Agricultura, Industria e Comercio (1917:256–336).

46 Reported by Roberto Santos as 13,282 contos de réis (1890), 65,390 contos (1900), and 75,625 contos (1910), converted respectively at 0.46 dollars per milréis, 0.19 dollars per milréis, and 0.33 dollars per milréis (see Santos, 1980:194).

TERRITORIAL CLAIMS AND CONCESSIONS

During the Rubber Boom, the longstanding geopolitical contest for Amazonia intensified as serious boundary conflicts erupted between Brazil and her neighbors over forest lands in the basin. Subject to competing colonial interests since the late fifteenth century, successive treaties divided the basin increasingly westward in favor of the Portuguese Crown, according to rights of possession *de facto* rather than *de jure* (Tambs, 1974). Dissolution of the Spanish American Empire blurred the boundaries separating the new republics, although each republic soon recognized the need to exercise dominion over their claims in the face of imperial Brazil. Peru, perhaps the most aggressive of the new republics, established a naval base at Iquitos in 1861 (Romero, 1983:21), signaling its intent to protect territorial claims and contest the Brazilian *Marcha para oeste* (Tambs, 1974). In addition, both Peru and Bolivia encouraged scientific exploration in the region, and sponsored colonization and immigration, particularly by Europeans, although neither succeeded in augmenting significantly the population of its Amazon territories (Fifer, 1972; Chirif, 1989).

The rise in rubber prices accomplished what no previous colonization program had managed to do: it drew tens of thousands of migrants into the basin, up from the east in Brazil and down from the Andean high jungles in search of rubber. Rubber workers penetrated the remotest reaches of the basin where the limits of activity were marked not by international boundaries but by hevea estradas and fallen caucho trees. Brazilian seringueiros settled in forest lands claimed by Bolivia and Peru, while the more transient Peruvian caucheros combed forests claimed by Brazil, Colombia, Ecuador, and Bolivia for the *castilloa* tree. The rich rubber fields of Acre became hotly contested, bringing Brazil to the brink of war in 1902 with Bolivia and two years later with Peru (Ganzart, 1934; Tambs, 1966; Rivière d'Arc, 1978). After several skirmishes and intensive negotiations, the conflict with Bolivia was resolved by a 1903 treaty in which Bolivia ceded nearly 200,000 square kilometers to Brazil in return for 5,460 square kilometers, a $10 million indemnity, and the promise of a railway around the cataracts of the Madeira River. The large capital expenditures made by the Brazilian federal government in Acre for the construction of the Madeira-Marmoré railway and restitution to Bolivia were offset almost completely by rubber and import revenues gathered in the region from 1903–1910 (LeCointe, 1922:II:413). The boundary dispute with Peru was settled by treaty in 1909, with Peru ceding some 400,000 square kilometers in the Upper Purus and Juruá region to Brazil in return for 39,000 square kilometers (Ganzart, 1934:438–439,447). Less acute conflicts also arose over the borders between Peru and Bolivia, Peru and Colombia, and Peru and Ecuador.

Thus the Upper Amazonian republics sought to establish and protect their claims to territory in the basin, and such efforts implied the expenditure of rubber revenues. Peru maintained a substantial naval fleet at Iquitos throughout the boom (Romero, 1983), and a sizable proportion of customs revenues was spent on supplying and maintaining the fleet. The budget proposed for the Department of Loreto for 1906, for example, allotted 56 percent of revenues expected from import and export duties to the military (Fuentes, 1908:I:271–287). Resolution of boundary conflicts also implied extraordinary expenditures of state funds, when armed forces were dispatched by Brazil, Bolivia, and Peru to the borders to protect or retake territory.

In addition, the central governments of several republics granted special navigation and land concessions to rubber entrepreneurs and firms for exceptionally large areas in the remote border lands to minimize the loss of rubber revenues and to assert territorial claims. In return for a concession, grantees typically would be required to provide transportation services to the region, sponsor colonization, establish public services, and maintain order. In Peru, cauchero C. F. Fitzcarrald was granted a transport monopoly on the Madre de Dios River a year before his death in 1897 (Reyna, 1942). Similarly, the British Peruvian Amazon Company was granted via Peruvian trader J. C. Arana, an extensive area and navigation rights in the disputed Putumayo region. In late 1905, the Colombian government granted Colombian Dr. L. Cuervo Márquez title to 400 square kilometers and a concession to a similar amount of forest (*U.S. Monthly Consular Reports*, no. 298, 1905 pp. 218–219). In Bolivia, the Barbo Contract of 1880, land concessions to N. Suárez and A. Vaca Diez, and the Bolivian Syndicate (Aramayo Contract) grant in Acre were all intended to buttress Bolivian hegemony in the region (Tambs, 1966:263,270). Such concessions may be considered as indirect subsidies from the state, sometimes supplemented by direct payments or loans. Although most concessions were granted for long periods, typically twenty-five years or more, few lasted the duration. In general, concessions were less successful in securing territorial claims than the more expensive strategy of maintaining a significant military presence in the region.

Regional Articulation and Administration

The advent of the rubber trade, enabled by free navigation and steam transportation on the Amazon River, furthered incorporation of Amazonia into the national and political spheres of the countries sharing the basin. Prior to the Rubber Boom, Amazonia was a vast and little-known region from which flowed a diversity of rather modest and peculiar riches. Petty traders carried medicinal plants, dyes, fibers, and waxes from the lowland rain forests, and Panama hats brought down from the high jungle towns of the eastern slopes of the Andes, propelled by oar and sail, to exchange their products for crude manufactured goods at Belém. Except for this port-city at the mouth of the Amazon, the Amazon River and its tributaries had neither urban centers, financial institutions, nor telecommunications—only minimal political authority and the barest of civil administrative apparatus.

With the growing promise of rubber in Amazonia, new opportunities and challenges emerged for the state: taxes had to be collected and redistributed, a variety of services provided to promote trade, and territory defined and protected from internal and external threats. Mounting rubber revenues brought calls for increased local autonomy, and secessionist movements sprang up across the basin that threatened national territorial claims and access to surplus from the trade. Although most revolts were urban-based (in Iquitos, Manaus, and Belém) and ultimately inconsequential, the Acre conflict began with Brazilian tappers declaring independence in Bolivian territory (Ganzart, 1934; Loureiro, 1986:123–127). A certain measure of autonomy was granted to the regions by the respective central governments (particularly in Peru and Brazil) by delegating powers of authority, taxation, and administration. Entire regional and local governments were established along with new states, departments, provinces, municipalities, and districts run by well-paid functionaires. Duties collected on exported rubber allowed regional administrations to borrow on foreign markets to finance periodic deficits and large capital projects, a practice that led to large external debts by the peak of the boom. For example, the state of Amazonas and the municipality of Manaus which had no public debt in 1892 owed some 61,087,160 Francs ($11,789,822) in 1902, and 132,408,333 Francs ($25,554,808) by 1913, of which 60 percent was due to foreign lenders (LeCointe, 1922:II:435–436). Rivalries developed between administrations within the region as each vied for importance and power in the rubber trade. Indeed, the construction of grandiose public buildings like the Manaus Opera Theater was partly a manifestation of the fervent contest that developed during the boom between administrations in Manaus and Belém.

STATE SUBSIDIES, GRANTS, AND DEVELOPMENT POLICIES

Between 1880 and 1910, Amazonia's impressive growth in public facilities and services included high-capacity port facilities, subsidized steamer routes covering tens of thousands of miles each year connecting the remotest rubber post with New York and Liverpool, and telegraph and wireless communications that linked the port-cities of the Amazon with their respective capital cities. Public investment was concentrated almost entirely within the urban areas of Belém, Manaus, and Iquitos. These cities flourished as vibrant social and economic centers, boasting among the era's finest utilities—waterworks, light and power, telephone, and tramways—as well as schools, hospitals, and public services to meet the needs of burgeoning populations engaged in the rubber trade and related activities (Cruz, 1967) (See Photos 14–17).

To stimulate the development of public facilities, regional administrations granted concessions by tender to private firms that would be responsible for completing the works and operating them for the period specified. Concessions were granted almost entirely to foreign firms for

Photo 14. Port of Belém, ca. 1907
(from Mattoso, 1908:212)

Photo 15. Rua 15 de Novembro, Belém, ca. 1907
(from Mattoso, 1908:328)

Photo 16. Theatro da Paz, Belém, ca. 1907
(from Mattoso, 1908:267)

Photo 17. Civic Water Reservoir, Belém, ca. 1907
(from Mattoso, 1908:265)

everything from the Manaus slaughterhouse, public markets, city water works and tramways to the Amazon telegraph cable and steamer transportation in the basin. Among the largest grants made were concessions to the British Manaos Harbour Ltd. and the U.S. Port of Para Co. which undertook the needed expansion of dock facilities and operation of port facilities under long term contracts. The Brazilian government retained the right to purchase the facilities before their reversion when the concession expired. Important subsidies also were provided throughout the boom to foreign and domestic shipping companies to support steamer service along the Amazon and its tributaries.

In addition to providing grants and subsidies to support the development of public facilities (used mostly by the rubber trade), the state also attempted to diversify the economy through public expenditure and investment. State governors and administrators were well aware of the growing dependence on the rubber sector during the boom. The state of Pará, the original locus of the rubber trade and the strongest seat of non-rubber interests in Amazonia, made several attempts to promote colonization, agricultural development, and industrialization in the Belém area (Weinstein, 1983b:110–123,92–94). The limited success of efforts to diversify the rural economy stemmed largely from the economic environment created by the rubber boom: the extractive sector drew immigrants away from the colonies to work in the more lucrative rubber trade, because agricultural estate owners could not afford the high wages demanded to retain labor. Industrial ventures supported by the state faced not only the labor scarcity problem but undercapitalization, high input costs, and low effective demand.

Fundamentally, the political economy of Amazonia during the era was shaped by the high returns in the rubber sector. Although government entities at various levels had access to substantial revenues captured from the trade which could have been directed toward diversifying the economy, powerful forces directed public investment into expanding infrastructure and facilities required by the rubber industry over other enterprises. The legitimacy of ascendant local state officials was constantly tested during the boom by rising expectations among the increasingly powerful rubber constituency for public investment to support the trade. As the boom developed, groups connected with the rubber industry assumed positions of power in local and regional administration, thus increasing pressure to invest public funds around the rubber trade. Allocating state revenues to other sectors in an effort to diversify the economy required using political capital that flowed increasingly from the economic power associated with the trade. The Lemos administration in Pará during the early 1900s cut off funding of projects begun by previous administrations to promote diversification and deliberately consolidated power around surplus captured from the rubber trade (Weinstein, 1983b:133–136). In the newer regions born of the rubber trade, such as Manaus and Iquitos, non-rubber interests were little developed prior to the boom, making the problem of local support for diversification efforts even more acute. The progressive centering of the regional political economy on the rubber trade further reinforced dependence on the extractive sector and heightened vulnerability to a fall in rubber prices.

After rubber prices plummeted and the region slid into economic crisis, the private sector turned to the state for urgently needed assistance and support. Although the state's efforts to support and reform the trade by such measures as the "National Campaign for the Defense of Rubber" (Weinstein, 1983b: 228) and to promote industrial diversification ultimately failed, the fact that the state was called on to remedy such a variety of local problems suggests that the state had become a major player in Amazonia during the boom. As the rubber trade diminished, the private sector receded, leaving the state in dominance in the region, a legacy that persists to this day in most Amazonian countries.

Chapter 10

THE RUBBER BOOM AND AMAZONIAN DEVELOPMENT: LESSONS AND LEGACIES

Introduction

Our analysis of the wild rubber industry (Part I) and of the macroeconomy of Amazonia during the Rubber Boom (Part II) are brought together in this chapter to provide a revised understanding of the boom and its legacies. We begin by offering a re-interpretation of the boom based on our findings. Of particular interest here is the joining of insights we garnered by building our analysis up from the microeconomics of wild rubber extraction to the macroeconomics of the boom economy. We turn then to consider the legacies of the boom, many of which are evident today and relevant to understanding the development path of Amazonia. Finally, we point to new directions for research to further enhance our understanding of this important era in Amazonian economic history.

The Rubber Boom Revisited

Previous accounts of the Rubber Boom and Amazonian economic development have pointed toward a failure in capital accumulation, arguing that surplus was drained by unequal exchange and foreign investors, that surplus creation and reinvestment were thwarted by inefficient pre-capitalist relations in wild rubber extraction, or that competitive production methods, in the form of rubber plantations, were limited by the environmental impediment of South American Leaf Blight. Our explanation shifts the focus to the geographic, market, and political factors that shaped the form of accumulation in such a way that substantial savings, investment, and growth occurred, greatly expanding the geographic scale of the regional economy but without fundamentally transforming its highly fragile and undiversified economic base.

The profitability of wild rubber extraction during the Rubber Boom led to a remarkable expansion in the industry, which began during the 1860s in the areas surrounding Belém and extended into the uppermost reaches of the Amazon Basin by the early 1900s. The associated investments in establishing estates, setting up tappers, developing information and transportation networks, and building transshipment facilities proved to be highly specific to rubber extraction either because of their inherent nature (as with estradas) or the lack of other extractive activities in these areas that could generate comparable returns. Furthermore, the linkages and technological spinoffs to other industries in the tradeables sector were not extensive. When rubber prices collapsed after 1910, the value and use of most of the investments in rubber extraction and transport dropped dramatically. Capital that could be mobilized was removed from the region in search of more profitable uses; much of the rest was effectively scrapped or put to less intensive use in extractive activities with much lower returns than rubber.

Investment beyond rubber extraction and transport was found largely in the non-tradeables sector. Prices and incentives of the Rubber Boom favored non-tradeables over other tradeable goods and few ancillary, self-propelling industries spun off from the rubber trade. The income needed to support the capital value of investments in real estate, urban construction, and service industries (and their potential as a source of wealth for financing structural adjustment after the boom) depended directly on the vitality of the tradeables sector as well as on the sector's ability to maintain regional incomes. Unfortunately, the viability of the few incipient import substitution industries that had developed (as in making paper, beer, soap, and packing) also depended on demand driven by income generated by rubber.

Basic competitiveness problems in the region, including the high costs of inputs and transportation, small market size, and high transaction costs were not solved by the boom. The drop in labor costs with the bust was not enough to improve industry's competitive position substantially in other tradeable sector activities, partly because much of the labor force migrated out or went into the hinterland in search of agricultural land. Agriculture proved to be a source of growth for the region in the wake of the boom, but its potential as an export sector was limited by the low level of investment made during the boom and the competitive disadvantages of growing basic food crops in the isolated Amazon Basin for interregional and international trade. Average incomes in agriculture remained well below those of the rubber era, and production was destined more for subsistence consumption than for sale. In the extractive sector, subsequent expansion of Brazil nut and other forest product exports provided revenues nowhere near the levels of rubber. With the tradeables sector far less vibrant than during the boom, the urban non-tradeables sector shrank dramatically, leaving the decay of urban splendors as a haunting reminder of the earlier boom.

State activity played a crucial role in shaping the region's development. Revenues captured by the state via major import and export taxes gave the state the financial potential to guide the path of the boom. Nevertheless, state actions tended to reinforce the logic of the private market, which was to push rubber extraction and transportation investments to the furthest frontiers and to deepen the non-tradeables sector. Territorial claims among the various Amazon countries and the ambitions of regional governments alike depended on establishing and extending economic activity throughout the basin. Securing territorial claims in frontier areas, articulating regional governments, and developing the administrative capacities to govern consumed much of the state's resources, especially for the smaller countries of the Upper Amazon, such as Peru and Bolivia, for whom Brazil's westward expansion posed a threat to their claims to vast low jungle areas. Both Peru and Bolivia lost significant areas on their eastern frontiers where rubber tappers, patrons, and traders tied to Brazil provided the basis for legitimizing the larger country's territorial claims. State consolidation and the expansion of rubber extraction during the boom were inextricably linked in Amazonia.

Abundant tax revenues derived from the rubber trade and the availability of foreign credit (leveraged by such revenues) allowed the forementioned nation-building goals to be advanced rapidly enough that the state was also able to take a significant role in shaping development policy through projects and major social expenditures. In the short run, state investments faced the same basic returns and incentives of private sector activities, and thus it is not surprising that public spending reinforced the dynamics of the boom economy by investing where returns were high, specifically in rubber extraction and in non-tradeables. Such an assertion, however, would slight the repeated efforts of state leaders to sponsor diversification projects, particularly in subsidizing incipient industries, and to search, often via public commissions and task-forces, for ways to create a more balanced regional economy. State development strategists and politicians were aware of the precarious nature of the boom, yet most of the projected revenues were spent on facilitating the expansion of rubber or promoting activity in the non-tradeables sector of urban construction and infrastructure development. Revenues were not built up for later disbursement, nor were price incentives dramatically altered to change the balance of sectoral activity. Instead, foreign credit allowed even more rapid investment in the booming sectors than otherwise would have been possible, so that by the time the crash came, foreign debt was substantial among the major rubber exporting states and (like many other investments in the region), fundamentally unbankable without a continued stream of rubber revenues. Further surplus retention from rubber exports during the boom would not necessarily have led Amazonia to a more sustainable development path because it was this very same surplus that distorted incentives for private and public investment and fostered the growth trajectory of a fragile and vulnerable economy.

Development Legacies of the Rubber Boom

Prior to the boom, only Belém was integrally linked to the international economy. But by the peak, few areas along tens of thousands of miles of rivers and even the vast interfluvial uplands had been left untouched. Floral and faunal resources had been affected substantially by settlement and intrusion. Native peoples had been driven deeper into the upland forests or incorporated into the burgeoning rubber trade. The landscape of the Amazon Basin had been transformed: estradas had been cut and land cleared for cultivating perennials and cattle along the Amazon and even on some rubber estates. Transportation and communications networks had been extended to the upper reaches of the more remote rivers. Major urban centers had been developed by investing substantial public and private resources in modern amenities. Property rights and territorial boundaries were defined in ways that proved to be rather durable. Administrative structures of the state and government authority had been established. In sum, the market and the state had become the dominant forces shaping the region's development and landscape, and their organization bore the stamp of rubber.

Several legacies were central to the organization and operation of the post-rubber economy. One was the establishment of the private estate as the dominant form of tenure throughout the basin, a legacy that persists today in much of the Brazilian Amazon where tappers now work their own estates or continue to tap rubber on the patron's property. In more marginal rubber producing areas, private estates provided the basis for developing a patron-tenant relationship similar to the land-labor interlinkage of hacienda agriculture elsewhere in Latin America. Tappers became tenants, and patrons granted access to land for agricultural and extractive activities, generally in a share-cropping relationship that sometimes involved the tenant providing direct labor to the patron. Private estates, especially in areas near the major urban centers, became the base for the region's post-rubber tradeables sector, providing a mixture of commercial agriculture (rice, sugar, cotton, coffee, and cacao) and extractive activity (Brazil nuts and rubber).

A second legacy was the extension of trading networks throughout the Amazon Basin. Even though the volume of trade to the Upper Amazon fell dramatically after the boom, in the far reaches of the Amazon Basin rubber continued to be tapped, Brazil nuts were gathered, and other products were extracted for export to foreign markets. Whereas the mix of economic activities on these distant estates came to include considerably more agriculture, hunting, and extraction for subsistence than at the peak of the boom, trade networks established in the previous era continued to link these areas to international markets and to integrate them to varying degrees into the development process in their respective countries. Areas once beyond the frontier were now integrated into national and world economies.

A related legacy was the form of trade relations that governed extractive activities along the river. The main contractual forms during the boom were barter trade among Amerindian indigenous communities and traders as well as debt-merchandise trade among traders, patrons, and tappers. These basic relations (and variants) have proved to be durable and pervasive forms of trade contracts, particularly in the more remote reaches of the basin (Padoch, 1987; McGrath, 1989). This durability can be traced to the continuing emphasis on extraction and both the risky nature and potentially high costs of monitoring these activities under any other type of contractual relation.

A fourth legacy was the pattern of rural and urban settlement in the region. During the boom, native peoples were often displaced, pushed away from upriver lowland areas and into the upland forests. Perhaps the most invasive and disruptive force was the rubber workers of Peru and Bolivia, who swept through the upland forests in search of caucho trees, displacing, harassing, sometimes murdering and sometimes incorporating native peoples into their efforts to gather rubber. The establishment of rubber estates based on the hevea tree and trade activities along rivers throughout the basin also affected native groups; adversely, by cutting off their access to traditional use areas and spreading immigrant diseases rapidly, and perhaps less adversely by providing access to steel tools, weapons and other foreign goods of high use value. In many areas, however, tribes were pushed by or retreated from the advance of the rubber trade into the very remotest reaches of the basin where they remain today.

For immigrants to rural areas of Amazonia, rubber estates defined the pattern of settlement during and after the boom. During the boom, population was highly dispersed, as tappers were separated by their estradas along the thousands of miles of riverfront properties. Along the main stem of the Amazon River and other tributaries where rubber trees were scarce, agricultural estates and small service communities were established. As the boom collapsed, people left the cities in search of land to work and many became tenants on agricultural estates or old rubber estates that had turned to other extractive products. In some areas, such as northeastern Peru, rubber estates begat loose communities and eventually villages, as tenants moved in closer to the patron, or the center of the estate, where they became farmers as well as tappers, hunters, and forest product collectors (Coomes, 1995). The location and organization of these villages, on the rubber estates, agricultural estates, and surrounding the secondary towns, are a direct legacy of the pattern of economic activity generated by the Rubber Boom.

The establishment of major urban centers, especially the jungle cities of Manaus and Iquitos, also is an important legacy of the boom. As the primary focus of investment in infrastructure for trade and almost the exclusive recipient of investment in the non-tradeables sector, the Rubber Boom created modern cities in the Amazon. Even with the declines they suffered following the collapse of the boom, Belém, Manaus, and Iquitos continue to play decisive roles in the region as service centers linking the surrounding rural areas to international markets for extractive goods, as direct sources of demand for agricultural and forest products from the region, as poles for future economic development projects, and as vents for public investment. In establishing these cities, the Rubber Boom created a legacy of urban bias in development programs that continues to this day, in at least the Upper Amazon.

Finally, the Rubber Boom firmly established the presence of the state throughout the basin. During the era, territorial rights were redefined, regional governments were created and articulated with central governments, and structures were set up for regional administration. The state became a major participant—as beneficiary, regulator, and investor—in forming the region's economy and thus in providing the basis for state-sponsored colonization and settlement projects and major infrastructural undertakings elsewhere in the region. What remains to be ascertained more systematically is how the particular state structures, ideologies, and groups that emerged during the Amazon Rubber Boom affected the subsequent evolution of the state and development in the region.

Toward an Improved Understanding of the Amazon Rubber Boom

Our original purpose in revisiting the Amazon Rubber Boom was to contribute to the constitution of a more satisfying interpretation of this particularly formative era in Amazonian economic history. Wide variations in the organization of the wild rubber industry during the boom were evident from contemporary accounts of the boom—from smallholders tapping a few estradas near the port-cities to the enormous estates of the rubber barons upriver—but could not be explained by prior interpretations of the era. In conducting a systematic examination of the organization, operation, and impacts of the wild rubber industry, we acquired new insights into the Amazon rubber industry at the end of the past century and elucidated the development logic of the boom. Some of these insights are relevant not only to the rubber era but also more broadly to ongoing discussions of resource-based, export-led growth and the organization of other extraction-based industries elsewhere.

Perhaps the most important conclusion of our work for students of Amazonian history is that the failure of the Rubber Boom to generate sustained economic development in the region lies in the nature of public and private investment of surplus generated by the boom and the fragile economic structures that these investment patterns engendered. Although in hindsight this finding seems rather painfully obvious, it stands in sharp contrast to previous explanations which argued that the failure of the boom was due to the rubber sector's inability to generate and retain adequate surplus or to modernize effectively. For many Latin American countries in the midst of economic recovery fueled in no small part by booming resource sectors and/or substantial capital inflows, the lesson we draw from the Rubber Boom is particularly salient: how surplus derived from such booms is managed and invested—particularly by the state—is crucial in conditioning future prospects for economic development. Unfortunately, the political economy problem of how to manage the surplus effectively, when social and economic pressures push public policy and expenditures toward aiding booming and non-tradeables sectors, remains and requires further exploration.

Of perhaps more value to policy analysts and researchers seeking to assess the development and conservation implications of other extractive activities is the methodological approach demonstrated in the microeconomic analysis of the Amazon wild rubber industry. In contrast to previous studies of wild rubber and many other extractive industries, we built our analysis from the "bottom-up," beginning with the basic physical features of wild rubber, the technology of its extraction and trade, the relative availability of the key productive factors used in the industry, and the region's competitive position relative to other wild rubber sources in a world market where demand for rubber was burgeoning. This micro-level approach highlighted the significant levels of transactions costs and risk present in wild rubber extraction and trade, which when combined with the rest of the industry-level analysis helped to reveal the logic of wild rubber's contractual and social relations. Of particular importance were the geographical variations in local biophysical and social conditions across the basin that gave rise to a variety of distinct relations of extraction and trade, which in turn conditioned the distribution of economic returns to participants and shaped patterns of investment. And, such investment by industry participants as well as the state conditioned prospects for longer-term development in the region. By articulating several levels of analysis, we were assured of a more comprehensive, integrative and coherent understanding of the organization and performance of the wild rubber industry that, in the end, allowed us to move beyond the prevailing explanations of the development experience associated with the boom.

Although comprehensive in scope of argumentation and novel in approach, our micro- and macroeconomic analyses reveal the need for further research to deepen our new found understanding of the Amazon Rubber Boom era. The study serves to set the stage for a second round of more focussed empirical research on the industry's organization, performance, and development experience. Of particular importance at the micro-level of analysis would be the task of specifying in some historical detail the nature and extent of spatial and temporal variations in the key biophysical and social conditions, the associated relations of production that arose for rubber extraction and trade, and their joint influence on the development experience of particular locales within the basin. Geographically specific studies are needed of the configuration of land holdings, natural resource and agricultural land use,[48] as well as the evolution of property, labor and capital relations during and after the Rubber Boom. Comparative analyses of the development experience during the boom across locales, such as upriver versus downriver areas, closed rivers versus open rivers, or caucho producing versus hevea producing areas, would be most informative. Such research would offer more *ex post* evidence on the outstanding question of the size and distribution of returns from rubber activities during the boom, and aid our understanding of the dynamics of the boom and the ensuing patterns of development.

---

48 Especially perennial crop production, see Coomes (1995).

At the macro-level of analysis, the research challenge lies in sorting out the reasons why and specifically how state policy fell into the Dutch Disease trap. The difficulty here lies in choosing among a rich variety of potential explanations. Regional competition among the urban centers of the rubber trade, particularly Manaus and Belém but also Iquitos, pushed municipal and state governments to improve infrastructure and the urban setting in order to attract or secure the heavy commercial activity associated with rubber. Restraining state expenditures on public development projects when revenues and foreign credit were abundant seems to have been nearly impossible for state officials facing pressure from a variety of constituencies to distribute the gains from the boom or to reinvest the gains in new endeavors. Moreover, the ascendant classes, especially in the new cities of Manaus and Iquitos, were those most closely tied to the dynamic sectors of the boom. State promotion of these activities, particularly given its huge fiscal resources, reinforced their positions and their wealth. The labor-intensive nature of construction, urban maintenance, public schools, and other urban infrastructural investments also made these activities popular with the burgeoning urban labor force. The alternative—setting aside earnings for a future bust, and slowing expansion of the burgeoning sectors via taxation and subsidization of other tradeables production—would have been singularly unpopular with all but the old-guard Paraense *fazendeiros*, planters, and any new producers of tradeable goods. A closer look at the historical evolution of the political economy of Amazonian states during this era would clarify the primary forces behind state policy formation, how such policy shaped regional economic structures, and why states failed to pursue alternate development paths.

Our analysis also points to a number of specific questions related to such themes that merit further empirical study. Again, many of the arguments made above were predominately conceptual in nature, drawn by deduction or by elimination, and supported by empirical information culled principally from secondary rather than primary sources. Additional research is required to assemble primary data from archival sources to address the following micro- and macro-level questions:

- What were the specific input costs, objective risks and conditions, and economic returns for the various participants (e.g., hevea tappers, caucho collectors, patrons, river traders, urban rubber merchants) in the Amazonian rubber trade at specific locations and times during the boom?

- To what degree were rivers of the basin open (or closed) to free trade over the fifty year history of the boom? What types of collusive arrangements might have emerged among traders? How did aviamento traders effectively police itinerant traders? How durable and effective was river closure? What proportion of rubber collected actually flowed down closed versus open rivers in Brazil?

- What was the experience and fate of the small estate holder (e.g., caboclo squatters, recent arrivees, etc.) during the boom, especially those near the port-cities? How did returns to such participants compare with those of tappers further upstream? Did contractual trade relations for holders living more than a day's travel from the city differ substantially from those much further upriver? When the boom collapsed, what happened to the small holders?
- How did participants in the rubber trade reinvest their earnings? What patterns of investment can be discerned over space and time by occupation or position in the trade? What proportion of investment was returned to rubber collection and exchange versus other productive activities (e.g., agriculture, industry, urban services, real estate, etc.)? How did participants shift their investments in response to dropping rubber prices in the 1910s and 20s? How useful were the investments made with rubber earnings in the post-boom economy?
- How did regional government expenditures evolve over the boom, and what were the implications of these expenditures for the structure of the economy both during and following the boom? What forces were behind state policy formation as the boom developed and how can we account for the failure of state efforts to avoid the Dutch disease trap?
- What impacts did the boom have on the economies and development experience of other regions in South America? How did investment of revenues captured by industry participants and the states of Brazil, Peru and Bolivia outside Amazonia shape conditions elsewhere? What benefits were realized by regions such as southern Brazil, Argentina or Uruguay that supplied foodstuffs and other goods imported into Amazonia during the boom?

This list of questions is by no means exhaustive but does serve to signal several key issues that flow from our analysis. Researchers determined to assemble data from primary sources thus may be better armed to focus their search and collection efforts. Of particular value would be notarial records, personal diaries, account books, and shipping records of tappers, patrons, river traders, and foreign firms directly involved in the rubber trade during the boom; formal property surveys conducted to register or contest rubber estates claims; and, government records of public revenues, expenditures, and returns on public investments. Primary data are by no means easily procured from regional sources and researchers seeking to pursue these topics face a formidable challenge, given the very uneven condition of archives in the Amazonian port-cities, the often scant records that have survived to the present, and the difficulties faced in securing information from private sources on an era for which local sentiments even today are rather mixed.[49] Our study of the Rubber Boom, nonetheless, suggests the potential importance in finding and analyzing primary information. Those who succeed may be richly rewarded through the new insights they would bring to our understanding of the rubber era and the early development path of Amazonia.

---

49 On several return visits to Peru, the second author has sought to find primary historical information on the rubber era from public and private sources in Iquitos with only limited success. We remain hopeful that researchers may be more successful in such endeavors in Manaus and Belém, both of which were larger urban centers during the boom.

Chapter 11

REFLECTIONS ON THE STUDY OF NATURAL RESOURCE AND DEVELOPMENT OUTCOMES

In undertaking our research on the Amazon Rubber Boom, and particularly in reviewing recent literature on the era, we were struck by a singular tendency among social scientists working on the era (as well as more recent extractive periods) to structure their accounts and analyses according to the "grand theories" of economic development. Armed with theories put forward in the 1950s to early 1980s by structuralists, neoclassical orthodoxy, dependency, and neo-Marxian analysts, researchers have sought ardently to place the role of resource extraction in the broader economic path by which countries and regions develop or underdevelop. In the case of the Rubber Boom, scholars typically have chosen non-neoclassical theories and applied them "from the top down"; that is, from broad theory to the specific case of wild rubber. For them, the boom experience serves as an instructive "instance" of underdevelopment or to suggest the supremacy of one theory over another.

Perhaps our most basic contribution in this volume is to provide an initial impetus to push discourse away from grand theories and toward more carefully constructed analyses of the microdynamics and macro-outcomes of extractive resource activities. Unlike prior analysts, we do not see the debate over whether economic development is thwarted or advanced by a region or nation's reliance on natural resource extraction (or more generally, on primary products such as agriculture) to be a particularly fruitful line of inquiry. In our study of wild rubber, we found all grand theories to be wanting when brought under the rather harsh light of "empirical reality," particularly with respect to the rich heterogeneity found in both underlying conditions and development outcomes. It was, in fact, our finding of marked spatial variation in the relations of production within the wild rubber industry across the Amazon basin that led us to question prevailing views of the boom built on grand theory. Moreover, as "grand" theories, none could bring together micro-level features with macro-level outcomes, or provide a persuasive logic for the organization of extractive activity and its link to development. For these reasons, we were dissatisfied with such theories and we remain skeptical of the potential they might hold for understanding place-specific development outcomes in contemporary extractive activities. Such a view is consistent with a broader consensus that has emerged recently in development economics.[50]

In our view, a far more promising challenge lies in understanding the the micro, meso, and macro development processes related to resource extraction that give rise to specific economic and environmental outcomes in particular places and moments. Indeed, one of our primary purposes in this volume was to develop and apply a theoretically-informed, rigorous but tractable analytical approach for the study of the wild rubber sector and related development outcomes in Amazonia. Our choices of the specific foci of analysis and the conceptual devices to be employed were driven primarily by a particular sense of the key problems that impeded our understanding of the era, the logic of the organization of the industry, and the causes of specific development outcomes. We arrived at such a sense through a careful reading of the first-hand accounts written by industry observers during the boom, field studies on the legacies and personal histories of descendants of the boom, and by contrasting the former with recent historical accounts by scholars on the era. The analytical framework we constructed brings together micro- and macroeconomic models to address the key questions we identified. On the microeconomic side, we modified the classic Structure-Conduct-Performance Model from the literature on industrial organization to incorporate the salient characteristics of risk and transaction costs, and then used the model to structure our inquiry into the wild rubber industry. On the macroeconomic side, the Dutch Disease theory of resource boom economies guided our effort to make sense of the development patterns of Amazonia during the boom era.

The approach we developed for understanding the wild rubber industry and the Amazon Rubber Boom economy could be used elsewhere to better explain the logic and performance of other extractive industries and thus to forge more effective development and conservation policies. As we suggest

---

[50] Although grand theories of development do continue to hold considerable interest in those social sciences where contention over their ideological content is of considerable import, such theories are no longer commonly deployed as analytical tools in development economics. For a sense of the position and role of grand theories in development economics, the interested reader should refer to Hirschman (1981:1–24), Sen (1983), and Bardhan (1988). One view of the demise of grand theory in economic thought would be as follows. During the 1950s, optimism abounded among development economists, especially structuralists (e.g., Lewis, 1954; Nurske, 1953; Prebisch, 1950; Rosenstein-Rodan, 1943) that state involvement in promoting investment in certain activities, especially in the industrial sector, could accelerate growth and development in low-income countries. Substantial state involvement in the economy, such as through import substitution industrialization, gave rise however to very mixed outcomes. At the same time, substantive conceptual and empirical critiques, from both more radical perspectives and neoclassical orthodoxy, raised serious doubts about the theories themselves. Practical experience combined with skepticism over theory has persuaded many development economists that grand theories of development are of rather limited value beyond highlighting the substantive issues they raise.

in Chapter Six, our study is most relevant to an extractive boom where one industry dominates the frontier economy of a developing country. We built our analysis from the "bottom up," from the characteristics of the industry through to the macroeconomic environment created by the boom. In conceptual terms, we showed how specific supply and demand conditions, when considered within the context of inherent risk and transaction costs, shaped the productive and contractual relations that came to define the wild rubber industry. In turn, such relations—specifically among labor, capital and land owners—strongly influenced the position of participants in the industry and therefore the level and distribution of financial returns received and risks experienced. How such returns were then invested by both private agents and public agencies—largely either plowed back into the rubber trade or sunk in non-tradeable investments—profoundly influenced the development outcomes associated with the rubber sector and the longer-term path of the regional economy of Amazonia. In this way, we linked the microfoundations of the wild rubber industry to the broader macroeconomic economy of Amazonia during the boom. Such micro- to macro-level articulation is essential to any attempt to build up from the ground level to an economy wide view of development, and constitutes, in our opinion, a useful guide to researchers interested in understanding other extractive industries under boom conditions.

When considering other extractive industries with our framework, researchers may well encounter other distinct foci for analysis depending on the basic characteristics both of the resource and its market. For example, unlike wild rubber, certain extractive or natural resources may be highly concentrated in prime reserves, and such concentration provides opportunities for economic agents to seek to gain control over these resources.[51] Preemptive control of prime resources for a firm or state can mean lower costs of production and, hence, provide the opportunity to earn more resource rents while discouraging further entrants by forcing them to rely on high cost reserve sites; in this way, firms or states can also create the opportunity to realize monopoly rents. Depending on the degree of concentration of the prime resource, the ability to appropriate the resource (in political and economic terms), and the costs inherent in gaining control and making the initial investments in its extraction and processing, much of the interesting "conduct" in an extractive industry may surround the issue of strategic competition among firms and states to control the rents accruing to prime resource reserves (Barham *et al.*, 1994). Such strategic issues were not relevant in our historical case because of wild rubber's widely dispersed nature, the high costs for monitoring extraction and transport, and the relative abundance of wild rubber in the Amazon rain forest. However, strategic issues would be particularly important for many other resource industries, especially those involving minerals and hydroelectric power sites. In a sense, our contribution to on-going research on the strategic dimensions

---

[51] Examples of prime reserve resources in different historical moments in the Americas would include bauxite, banana lands, and hydroelectric sites (see Barham *et al.*, 1994).

of extractive industries lies in placing the wild rubber industry at the far end of a strategy-relevance continuum, and to highlight the potentially important role of risk and transaction costs in influencing industrial organization at both ends of this continuum.

Our approach to understanding the wild rubber industry, of course, would have been quite different had the rubber sector not dominated the Amazonian regional economy and given rise to an expansive boom. In a sense, our analytical task was simplified by the predominance of the industry in the regional economy. With no sector or activity offering competitive returns or trade volumes as wild rubber, we could focus our attention exclusively on the industrial organization and performance of rubber extraction and trade. It was not necessary to carefully situate the sector in the context of the tradeoffs facing individuals or households choosing to tap rubber over, say, the pursuit of other forest-related extractive activities, such as hunting, gathering fruit, or fishing. Wild rubber—and the non-tradeable sector built on this boom commodity—was the boom activity that drew people out of other sectors. However, where an extractive industry does not drive an economic boom—as is the case for most forest extraction products today (including wild rubber) in Amazonia—the challenge of understanding development outcomes becomes more formidable. Not only must we understand the organization of the specific industry, but we must also consider the conditions that influence decisions taken local residents over participation in the industry's activities.

Any prospective framework developed to analyze non-boom extractive activities would have to integrate considerations over participation into the analysis of the structure, conduct and performance of the industry. Potential participants in such activities face a variety of options, and with no obvious sector in dominance, they would tradeoff potential financial returns and risks in choosing to become involved in one or more activity. Among the key decisions made are whether or not to participate in the particular extractive activity, and the extent of participation relative to other activities (e.g., agriculture or wage labor). Such decisions shape the extent to which the extractive activity is pursued widely, which in turn affects the size and relationship of that sector to other regional economic activities (e.g., agriculture, wage labor, other extractive industries, etc.). The tradeoffs made by participants (i.e., both in terms of foregoing alternate economic activities as well as the potential spillover effects of the activity on other economic opportunities) would have to be accounted for in order to relate the sector's performance to broader development outcomes. Simply put, whereas during a resource boom there is essentially "only one game in town," during non-boom times, people must chose among a variety of games. Understanding the logic and development outcomes of extractive activity thus becomes a balancing act of studying economic sectors and household sectoral choices. Because sectoral and household outcomes are interdependent, they must be studied jointly.

Whether natural resource extraction constitutes a boom-time activity or acts more as a complement to other economic pursuits, our bottom-up approach will be useful in structuring subsequent analyses not only of the economics of extraction, but also of related issues that received only limited attention in our account. Our study of wild rubber focussed preferentially on improving our understanding of the micro- and macroeconomics of extraction, what for us were the most significant lacunae in the literature on the boom. As a result, we paid little attention to issues related to regional politics, class formation and behavior, ethnic relations during the boom, and cultural change brought by the boom—all issues of much relevance to a more complete understanding of the historical experience of the Rubber Boom and of much concern in previous works on the era. Their absence is perhaps most obvious in our treatment of the political economy of Amazonia during the boom, particularly with respect to understanding the specific difficulties experienced by state decision-makers over pursuing policies that might have served to diversify the economic base of the region or to set aside surplus in extra-regional assets for later use in the economy. With more emphasis on the types of class and sectoral alliances that were created in the Amazonian economy among various interests, and how they played out in the political arena of regional governments, we would have been better positioned to offer more historically specific arguments about the challenge of avoiding distorted development outcomes in boom economies.

In our view, considerable promise for developing a more comprehensive political economy of regional development in frontier areas such as Amazonia lies in building upon a firmer foundation of understanding the microeconomics of resource extraction industries and the macroeconomic conditions they create. Indeed, certain political, social and even cultural phenomena related to extractive industries may be studied more effectively by relating them explicitly to our bottom-up microeconomic approach rather than by considering them quite separately from the material conditions upon which such industries are built and the economic processes by which they are governed. Clearly, the study of issues such as class and ethnic relations would also improve our understanding of industry outcomes.

The prospects of exploring linkages that may exist between the microeconomics of resource extraction and regional politics, social conditions and cultural relations excites us. We are reminded of the lessons we have learned from one another, working together as economist and geographer, that informed our approach to better understand the Amazon Rubber Boom. By looking for ways to connect our respective disciplinary "trails through the forest," we found ourselves challenged to push further to better incorporate and integrate issues of mutual concern. We look forward to efforts of other scholars who may feel it worthwhile to seek to join their interests and approaches to our work in the larger project of better understanding the role of natural resource extraction in economic development. Ironically, such a project may lead even further away from grand theories of development and yet leave us with a much richer understanding of the diversity of the development experience. We are optimistic that such endeavors will bring rewards well worth the investment. That, at least, has been our experience thus far.

Bibliography

AIDESEP, 1991. "Esclavitud indígena en la región Atalaya," *Amazonía Indígena*, Año 11, no. 17&18 (September), pp. 3–13.

Akers, C. E., 1912. *Report on the Amazon Valley. Its Rubber Industry and Other Resources*. Waterlow and Sons Limited, London.

Alvim, Paulo de T., 1979. Agricultural production potential of the Amazon region. in: *Pasture Production in Acid Soils of the Tropics*. P. A. Sánchez and L. E. Tergas (eds.), Centro Internacional de Agricultura Tropical, Cali, pp. 13–23.

Anderson, Anthony B. and Edviges M. Ioris, 1992. "Valuing the rain forest: economic strategies by small scale forest extractivists in the Amazon estuary (Combu Island)," *Human Ecology*, vol. 20. pp. 337–369.

Anderson, A. B. and M. A. Jardim, 1989. "Costs and benefits of floodplain management by rural inhabitants in the Amazon estuary: a case study of açai palm production," in: *Fragile Lands of Latin America: Strategies for Sustainable Development*. J. O. Browder (ed.), Westview Press, Boulder, CO., pp. 114–126.

Anderson, A. B., P. H. May, and M. J. Balick, 1991. *The Subsidy from Nature: Palm Forests, Peasantry, and Development on an Amazon Frontier*. Columbia University Press, New York.

Arnous de Rivière, Baron H., 1900. "Explorations in the rubber districts of Bolivia," *Journal of the American Geographical Society of New York*, vol. 32, pp. 432–440.

Auty, Richard M., 1993. *Sustaining Development in Mineral Economies: The Resource Curse*. Routledge, New York.

Bakx, Keith, 1988. "From proletarian to peasant: rural transformation in the State of Acre, 1870–1986," *The Journal of Development Studies*, vol. 24, no. 2, pp. 141–160.

Ballivián, Manuel V., and Casto F. Pinilla, 1912. *Monografía de la Industria de la Goma Elástica en Bolivia*. Dirección General de Estadística y Estudios Geográficos, La Paz.

Bardhan, Pranab K., 1988. "Alternative approaches to development economics," in: *Handbook of Development Economics*, vol. 1. H. Chenery and T. R. Srinivasan (eds.), Elsevier Publishers, London, pp. 39–71.

________. (ed.), 1989. *The Economic Theory of Agrarian Institutions.* Clarendon Press, Oxford.

Barham, Bradford L., 1988. "Excess capacity in international markets: an application to the banana industry in Latin America". Ph.D. Dissertation, Stanford University, Stanford, CA.

________. 1994. "Strategic capacity investments and the Alcoa-Alcan monopoly: 1888–1945," in: *States, Firms, and Markets: The World Economy and Ecology of Aluminum*, Bradford Barham, Stephen G. Bunker, and Denis O'Hearn (eds.). University of Wisconsin Press, Madison, WI, pp. 69–110.

Barham, Bradford, nd. "Strategic committments in scarce resource industries: microfoundations of multinational monopolies" (unpublished manuscript).

Barham, Bradford L., and Oliver T. Coomes, 1994a. "Reinterpreting the Amazon rubber boom: investment, the state, and Dutch disease," *Latin American Research Review*, vol. 29, no. 2, pp. 73–109.

________. 1994b. "Wild rubber: industrial organisation and the microeconomics of extraction during the Amazon rubber boom (1860–1920)," *Journal of Latin American Studies*, vol. 26, no. 1, pp. 37–72.

Barham, Bradford, Stephen G. Bunker, and Denis O'Hearn, 1994. "Raw material industries in resource-rich regions," in: *States, Firms, and Markets: The World Economy and Ecology of Aluminum*, Bradford Barham, Stephen Bunker, and Denis O'Hearn (eds.). University of Wisconsin Press, Madison, WI, pp. 3–38.

Baumol, William J., John C. Panzer, and Robert D. Willig, 1982. *Contestable Markets and the Theory of Industry Structure.* Harcourt, Bracem Jovanovich Inc., New York.

Bonilla, Heraclio, 1977. "El caucho y la economía del oriente peruano," in: *Gran Bretaña y el Perú: los mecanismos de un control económico.* vol. V, Instituto de Estudios Peruanos, Fondo del Libro del Banco Industrial del Perú, Lima, pp. 123–133.

Bonilla, Heraclio, 1976. *Gran Bretaña y el Perú. Informes de los cónsules: Callao, Iquitos y Lambayeque (1867–1914).* Instituto de Estudios Peruanos, Fondo del Libro del Banco Industrial del Perú, Lima.

Bottomley, A., 1975. "Interest rate determination in underdeveloped rural areas," *American Journal of Agricultural Economics*, vol. 57, pp. 279–291.

Bunker, Stephen G., 1984. "Modes of extraction, unequal exchange, and the progressive underdevelopment of an extreme periphery: the Brazilian Amazon, 1600–1980," *American Journal of Sociology*, vol. 89, no. 5, pp. 1017–1064.

Bunker, Stephen G., 1985. *Underdeveloping the Amazon. Extraction, Unequal Exchange, and the Failure of the Modern State*. University of Illinois Press, Chicago, IL.

Burkhalter, S. Brian and Robert F. Murphy, 1989. "Tappers and sappers: rubber, gold and money among the Mundurucú," *American Ethnologist*, vol. 16, no. 1, pp. 100–116.

Burns, E. Bradford, 1965. "Manaus 1910: portrait of a boom town," *Journal of Inter-American Studies*, vol. 7, no. 3, pp. 400–421.

Carleton, Dennis W. and Jeffrey M. Perloff, 1990. *Modern Industrial Organization*. Scott Foresman/Little, Brown Higher Education, Glenview, IL.

Chew, Tek-Ann, 1991. "Share contracts in Malayasian rubber smallholdings," *Land Economics*, vol. 67, no. 1, pp. 85–98.

Chibnik, Michael. 1994. *Risky Rivers: The Economics and Politics of Floodplain Farming in Amazonia*. The University of Arizona Press, Tucson, AZ.

Chirif, Alberto, 1989. "Poblaciones humanas y desarrollo amazónico: el caso del Perú," in: *Poplaçoes Humanas E Desenvolvimento Amazônico*. Universidade Federal do Pará, Belém, pp. 267–311.

Chirif, Alberto and Carlos Mora, 1980. "La amazonia peruana," in: *Historía del Perú*. Tomo XII, Editorial Juan Mejía Baca, Lima, pp. 219–321.

Church, Col. G.E., 1904. "The Acre territory and the caoutchouc region of south-western Amazonia," *Geographical Journal*, vol. 23, no. 5, pp. 596–613.

Coase, Ronald H., 1937. "The nature of the firm," *Economica*, November, pp. 386–405.

Coates, Austin, 1987. *The Commerce in Rubber: The First 250 Years*. Oxford University Press, New York.

Collier, Richard, 1968. *The River that God Forgot: The Story of the Amazon Rubber Boom*. Collins, London.

Coomes, Oliver T., 1992. "Making a living in the Amazon rain forest: peasants, land, and economy in the Tahuayo River basin of northeastern Peru." Ph.D. Dissertation, University of Wisconsin-Madison, Madison, WI.

————. 1995. "A century of rain forest use in western Amazonia: lessons for extraction-based conservation of tropical forests," *Forest and Conservation History*, vol. 39, no. 3, pp. 108–120.

________. nd. "Income formation among Amazonian peasant households in northeastern Peru: empirical observations and implications for market-oriented conservation" (unpublished manuscript).

Coomes, Oliver T., and Bradford L. Barham, 1994. "The Amazon rubber boom: labor control, resistance, and failed plantation development revisited," *Hispanic American Historical Review*, vol. 72, no. 2, pp. 231–257.

Cooper, Clayton S., 1917. *The Brazilians and their Country*. Frederick A. Stokes Co., New York.

Corden, W.M., and J.P. Neary, 1982. "Booming sector and deindustrialization in a small open economy," *The Economic Journal*, vol. 92, pp. 825–848.

Cruz, Ernesto, 1967. *As Obras Públias do Pará.* Vols. I & II, Imprensa Oficial, Belém.

Cruz, Oswaldo Gonçalves, 1972. "Relatório sobre as condiçoes sanitárias do vale do Amazonas," in: *Sobre o Saneamento da Amazônia.* [1913], Philippe Daou e Fundação Getúlio Vargas, Rio de Janeiro, pp. 45–155.

D'Ans, André-Marcel, 1982. *L'Amazonie Péruvienne Indigène. Anthropologie, Ecologique, Ethnohistoire, Perspectives Contemporaines.* Payot, Paris.

Dean, Warren, 1987. *Brazil and the Struggle for Rubber: A Study in Environmental History.* Cambridge University Press, New York.

DeKalb, Courtenay, 1890. "The business of rubber gathering in the Amazon valley, and its development by American capitalists," *India Rubber World and Electrical Trades Review*, vol. 2, no. 3 (June 15), pp. 191–193.

De Mello Petey, Beatriz Cécilia C., 1972. "Aspecto da economía Amazônica na epoca da depressão (1920–1940)," *Boletim Geográfico.* (Rio de Janeiro), vol. 31, no. 229, pp. 112–131.

Denevan, William M., 1992. "The aboriginal population of Amazonia," in: *The Native Population of the Americas.* W. M. Denevan (ed.), Second Edition, University of Wisconsin Press, Madison, pp. 205–234.

Dixit, Avinash K., 1980. "A model of duopoly suggesting a theory of entry barriers," *Bell Journal of Economics*, vol. 10, no. 1, pp. 20–32.

Domínguez, Camilo and Augusto Gómez, 1990. *La Economía Extractiva en la Amazonía Colombiana, 1850–1930.* Tropenbos Colombia, Corporación Colombiana Para la Amazonía Araracuara, Bogotá.

Edwards, William E., 1847. *A Voyage up the River Amazon, Including a Residence at Pará.* John Murray, London.

Enock, C.R., 1910. *The Andes and the Amazon: Life and Travel in Peru*. 4th imp. T. F. Urwin, London.

Falção, Emilio (ed.), 1907. *Album do Rio Acre, 1906–1907*. Anuario Comercial, Lisbon.

Fearnside, Philip M., 1989. "Extractive reserves in Brazilian Amazon," *BioScience*, vol. 39, no. 6, pp. 387–393.

Ferreira de Castro, José M., 1934. *Jungle, A Tale of the Amazon Rubber-Tappers*. Charles Duff (trans.), The Viking Press, New York.

Fifer, J. Valerie, 1970. "The empire builders: a history of the Bolivian rubber boom and the rise of the house of Suárez," *Journal of Latin American Studies*, vol. 2, no. 2, pp. 113–146.

________. 1972. *Bolivia: Land, Location, and Politics since 1825*. Cambridge University Press, London.

Flores Marín, José Antonio, 1987. *La Exploitación del Caucho en el Perú*. Consejo Nacional de Ciencia y Tecnología, Lima.

Fuentes, Hildebrando, 1908. *Loreto. Apuntes Geográficos, Históricos, Estadísticos, Políticos y Sociales*. Tomo I & II, Imprenta de la Revista, Lima.

Fuller, Stuart J., 1912. "Trade in Peruvian rubber district," *U.S. Daily Consular and Trade Reports*, no. 274 (Nov. 20), pp. 913–927.

Furtado, Celso, 1963. *The Economic Growth of Brazil. A Survey from Colonial to Modern Times*. University of California Press, Berkeley.

Ganzart, Frederic William, 1934. "The boundary controversy in the upper Amazon between Brazil, Bolivia, and Peru, 1903–1909," *The Hispanic American Historical Review*, vol. 14, no. 4, pp. 427–449.

Garnier, Lyonel, 1902. "Caucho gathering on the upper Amazon," *India Rubber World*, vol. 26, no. 3 (June 1), pp. 279–81.

Gilbert, Richard J., 1986. "Pre-emptive competition," in: *New Developments in the Analysis of Market Structures*. Joseph E. Stiglitz and G. Frank Mathewson (eds.), MIT Press, Cambridge. pp. 90–123.

Glade, William, 1989. "Economy, 1870–1914," in: *Latin America: Economy and Society, 1870–1930*. Leslie Bethell (ed.), Cambridge University Press, Cambridge, pp. 1–56.

Gonçalves, A.C. Lopes, 1904. *The Amazon. Historical, Chorographical and Statistical Outline up to the Year 1903*. 1st Edition, Hugo J. Hanf, New York.

Gray, Andrew, nd. "The Putumayo atrocities re-examined," (unpublished manuscript).

Grubb, Kenneth G., 1930. *Amazon and Andes*. The Dial Press, New York.

Hale, Albert, 1913. "Developing the Amazon valley," *Bulletin of the Pan American Union*, vol. 36, pp. 38–47.

Hardenburg, W.E., 1912. *The Putumayo, The Devil's Paradise*. T. F. Unwin, London.

Haring, Rita, 1986. "Burguesía regional de la Amazonía Peruana," *Amazonía Peruana*, vol. 7, pp. 67–85.

Hayami, Yujiro and Vernon W. Ruttan. 1985. *Agricultural Development: An International Perspective*. Johns Hopkins University Press, Baltimore.

Hecht, S.B., A.B. Anderson, and P. May, 1988. "The subsidy from nature: shifting cultivation, successional palm forests, and rural development," *Human Organization*, vol. 47, no. 1, pp. 25–35.

Hecht, Susanna and Alexander Cockburn, 1989. *The Fate of the Forest. Developers, Destroyers and Defenders of the Amazon*. Verso, New York.

Hemming, John, 1987. *Amazon Frontier. The Defeat of the Brazilian Indians*. Harvard University Press, Cambridge, MA.

Herrera, Genaro E., 1904. "Clima de la Amazonía," *Boletín de la Sociedad Geográfica de Lima*, Tomo XV, Trim. I, pp. 87–93.

________. 1914. "Censo urbano de Iquitos," *Boletín de la Sociedad Geográfica de Lima,* Tomo XXX, Trim. I and II, pp. 43–51.

Higbee, Edward C., 1951. "Of man and the Amazon," *Geographical Review*, vol. 41, no. 3, pp. 401–420.

Hiraoka, Mário, 1985a. "Changing floodplain livelihood patterns in the Peruvian Amazon," *Tsukuba Studies in Human Geography*, no. 9. Tsukuba, Japan: The University of Tsukuba, Tsukuba, Japan, pp.243–275.

________. 1985b. "Floodplain farming in the Peruvian Amazon," *Geographical Review of Japan*, vol. 58 [Series B], no. 1, pp. 1–23.

________. 1985c. "Mestizo subsistence in riparian Amazonia," *National Geographic Research*, vol. 1, no. 2, pp. 236–246.

Hirschman, Albert O., 1981. *Essays in Trespassing: Economics to Politics and Beyond*. Cambridge University Press, Cambridge, MA.

House of Commons Sessional Papers, Session of February 14, 1912 to March 7, 1913, vol. 68, Misc. no. 8., His Majesty's Stationery Office, London.

India Rubber World, (various issues), The India Rubber Publishing Co., New York.

India Rubber World and Electrical Trades Review, (various issues), The India Rubber Publishing Co., New York.

Iribertegui, Ramón, 1987. *Amazonas. El Hombre y el Caucho*. Vicariato Apostólico de Puerto Ayacucho, Monografía no. 4, Caracas.

James, Preston, 1930. "The Tapajóz and Xingú valleys of Brazil. A type study in the evolution of Amazon landscape," *Bulletin of the Geographical Society of Philadelphia*, vol. 28, pp. 62–77.

Jumelle, Henri, 1903. *Les Plantes à Caoutchouc et à Gutta. Exploitation, Culture et Commerce dans tous les Pays Chauds*. Augustin Challamel, Paris.

Katzman, Martin T., 1976. "Paradoxes of Amazonian development in a 'resource-starved' world," *The Journal of Developing Areas*, vol. 10, pp. 445–460.

________. 1987. "Review article: ecology, natural resources, and economic growth: underdeveloping the Amazon," *Economic Development and Cultural Change*, vol. 35, no. 2, pp. 425–437.

Labroy, M.O. and M.V. Cayla, 1913. *Culture et exploitation du caoutchouc au Brésil*. Société Générale d'Impression, Paris.

Lange, Algot, 1911. "The rubber workers of the Amazon," *Bulletin of the American Geographical Society*, vol. 62, pp. 33–36.

________. 1912. *In the Amazon Jungle*. G.P. Putnam's Sons, New York.

________. 1914. *The Lower Amazon*, G.P. Putnam's Sons, New York.

LaRue, Carl D., 1926. *The Hevea Tree in the Amazon Valley*. U.S. Department of Agriculture Bulletin no. 1422, Washington, D.C.

Lawrence, James C., 1931. *The World's Struggle with Rubber*. Harper and Bros., New York.

LeCointe, Paul, 1922. *L'Amazonie Brésilienne*. Vols. I & II, Augustin Challamel, Paris.

Leff, Nathaniel, 1973. "Tropical trade and development in the nineteenth century: the Brazilian experience," *Journal of Political Economy*, vol. 81, no. 3, pp. 678–696.

Levin, Jonathan, 1960. *The Export Economies: Their Pattern of Development in Historical Perspective*. Harvard University Press, Cambridge, MA.

Lewis, W. Arthur, 1954. "Economic development with unlimited supplies of labor," *Manchester School of Economics and Social Studies*, vol. 22, pp. 139–191.

Little, Ian M.D, Richard N. Cooper, W. Max Corden, and Sarath Rajapatirana, 1993. *Boom, Crisis, and Adjustment: The Macroeconomic Experience of Developing Countries.* Oxford University Press, New York.

Loureiro, Antônio José S., 1986. *A Grande Crise (1908–1916).* T. Loureiro & Cía., Manaus.

Mattoso, Ernesto, 1908. *Album do Estado do Pará.* Imprimerie Chaponet (Jean Cussac), Paris.

Maúrtua, Anibal, 1911. *Geografía Económica del Departamento de Loreto.* Litografía Tip. Carlos Fabri, Lima.

McGrath, David Gibbs, 1989. "The Paraense traders: small-scale, long distance trade in the Brazilian Amazon," Ph.D. dissertation, University of Wisconsin-Madison.

Melby, J.F., 1942. "Rubber river: an account of the rise and collapse of the Amazon boom," *Hispanic American Historical Review*, vol. 22, pp. 452–469.

Ministerio da Agricultura, Industria e Commercio, 1917. *Annuario Estatistico do Brazil (1908–1912*). Vol. II (Economia e Finanças), Directoria Geral de Estatistica, Tipographia da Estatistica, Rio de Janeiro.

Muratorio, Blanca, 1991. *The Life and Times of Grandfather Alonso: Culture and History in the Upper Amazon.* Rutgers University Press, New Brunswick, NJ.

Murphy, R. and J. Steward, 1956. "Tappers and trappers: parallel process in acculturation," *Economic Development and Cultural Change*, vol. 4, no. 4, pp. 335–353.

Nery, F.J. Baron de Santa Anna, 1901. *The Land of the Amazons.* George Humphery (trans.), Sands and Co. New York.

Netto, F. Ferreira, 1945. "The problem of the Amazon," *Scientific Monthly*, vol. 61, pp. 33–44 & 90–100.

Norgaard, Richard B., 1981. "Sociosystem and ecosystem coevolution in the Amazon," *Journal of Environmental Economics and Management*, vol. 8, no. 3, pp. 238–254.

Nurske, Ragnar, 1953. *Problems of Capital Formation in Underdeveloped Countries.* Oxford University Press, New York.

Ordinaire, Olivier, 1892. *Du Pacifique á l'Atlantique par les Andes péruviennes et l'Amazone.* E.Plon, Nourrit et Cie., Paris.

Oyague y Calderón, Carlos, 1913. "Contribución al estudio de la crisis del caucho en el Amazonas," *Boletín de la Sociedad Geográfica de Lima*, Tomo XXIX, Trim. I–II, pp. 176–213.

Pacheco de Oliveira Filho, João. 1979. "O caboclo e o brabo: notas sobre duas modalidades de força-de-trabalho na expansão da fronteira Amazônica no século XIX," *Encontros com a Civilização Brasileira*, vol. 11, pp. 101–40.

Padoch, Christine, 1988. "The economic importance and marketing of forest and fallow products in the Iquitos region," in: *Swidden-Fallow Agroforestry in the Peruvian Amazon*. William M. Denevan and Christine Padoch (eds.), Advances in Economic Botany, vol. 5, The New York Botanical Garden, Bronx, NY., pp. 74–89.

Padoch, C., 1987. "Risky business," *Natural History*, vol. 96, no. 10, pp. 56–65.

Padoch, C., J. Chota Inuma, W. de Jong, and J. Unruh, 1985. "Amazonian agroforestry: a market-oriented system in Peru," *Agroforestry Systems*, vol. 3, pp. 47–58.

Parker, Eugene, Darrell Posey, John Frechione, and Luiz Francelino da Silva, 1983. "Resource exploitation in Amazonia: ethnoecological examples from four populations," *Annals of Carnegie Museum*, vol. 52, pp. 163–203.

Pearson, Henry C., 1901. "The crude rubber supply," *India Rubber World*, vol. 23, no. 5 (Feburary 1), pp. 135–36.

Pearson, Henry C., 1911. *The Rubber Country of the Amazon*. The India Rubber World, New York.

Peck, Merton J. (ed.), 1988. *The World Aluminum Industry in a Changing Energy Era*. Resources for the Future, Baltimore.

Pennano, Guido, 1988. *La Economía del Caucho*. Centro de Estudios Teológicos de la Amazonía (CETA), Iquitos.

Pierrot, A. Ogden, 1924. "The Brazil-nut or Castanha industry," *Supplement to Commerce Reports*, Trade Information Bulletin no. 259, Foodstuffs Division, United States Department of Commerce, Government Printing Office, Washington, D. C.

Pinedo-Vásquez, Miguel, Daniel Zarin, Peter Jipp, and Jomber Chota-Inuma, 1990. "Use-values of tree species in a communal forest reserve in northeast Peru," *Conservation Biology*, vol. 4, no. 4, pp. 405–416.

Plane, Auguste, 1903. *A Travers l'Amérique Equatoriale. Le Pérou*. Plon Nourrit et Cie., Paris.

Posey, Darrell A. and William Balée., eds., 1989. *Resource Management in Amazonia: Indigenous and Folk Strategies*. Advances in Economic Botany, vol. 7, The New York Botanical Garden, The Bronx, New York.

Prebish, Raul, 1950. *The Economic Development of Latin America and its Principal Problems*. Department of Economic Affairs, United Nations, Lake Success, NY.

Rands, R.D., 1924. *South American Leaf Disease of Pará Rubber*. U.S. Department of Agriculture Bulletin no. 1286, Washington, D.C.

Rands, R.D. and Loren G. Polhamus, 1955. "Progress report on the cooperative Hevea rubber development in Latin America," *USDA Circular no. 976*, U.S. Department of Agriculture, Washington, D. C.

Redford, Kent H. and Christine Padoch, eds., 1992. *Conservation of Neotropical Forests: Working from Traditional Resource Use*. Columbia University Press, New York.

Resor, Randolph R., 1977. "Rubber in Brazil: dominance and collapse, 1876–1945," *Business History Review*, vol. 51, no. 3, pp. 341–366.

Reyna, Ernesto, 1942. *Fitzcarrald, el Rey del Caucho*. Taller Gráfico de P. Barrantes C., Lima.

Rivière d'Arc, Hélène, 1978. "La formation du lieu Amazonie au XIXe siècle," *Cahiers des Amériques Latines*, (Série Science de l'Homme), vol. 18, pp. 183–213.

Romanoff, Steven, 1992. "Food and debt among rubber tappers in the Bolivian Amazon," *Human Organization*, vol. 51, no. 2, pp. 122–135.

Romero, Fernando, 1983. *Iquitos y la Fuerza Naval de la Amazonía (1830 1933)*. 3rd Edition, Dirección General de Intereses Marítimos, Ministerio de Marina, Lima.

Rosenstein-Rodan, Paul, 1943. "Problems of industrialization of eastern and southeastern Europe," *Economic Journal*, vol. 53, no. 1, pp. 202–211.

Ross, Eric B., 1978. "The evolution of the Amazon peasantry," *Journal of Latin American Studies*, vol. 10, no. 2, pp. 193–218.

Russan, Ashmore, 1902. "Working rubber estates in the Amazon," *India Rubber World*, vol. 27, no. 1 (October 1), pp. 5–7.

Salamanca, Demetrio T., 1916. *La Amazonía Colombiana. Estudio Geográfico, Histórico y Jurídico en Defensa del Territorial de Colombia*. Imprenta Nacional, Bogotá.

Sánchez, Pedro A., 1981. "Soils of the humid tropics," in: *Blowing in the Wind: Deforestation and Long–Range Implications*. Vinson H. Sutlive, Nathan Altshuler, Mario D. Zamora (eds.), Studies in Third World Societies, Publication no. 14, Department of Anthropology, College of William and Mary, Williamsburg, VA., pp. 347–410.

San Román, Jesús V., 1975. *Perfiles Históricos de la Amazonía Peruana*. Centro de Estudios Teológicos de la Amazonía (CETA), Ediciones Paulinas, Lima.

Santos, Roberto, 1980. *História Econômica da Amazônia (1800–1920).* T.A. Queiróz, São Paulo.

Santos, Roberto Araújo de Oliveira, 1968. "O equilíbrio da firma 'aviadora' e a significaçao econômico-institucional do 'aviamento'," *Pará Desenvolvimento*, vol. 3, IDESP, Belém.

Scherer, Frederic M., 1980. *Industrial Market Structure and Economic Performance.* Third Edition, Rand McNally College Publishers, Chicago.

Schidrowitz, Philip, 1911. *Rubber.* Methuen and Co. Ltd, London.

Schmink, Marianne and Charles Wood, 1992. *Contested Frontiers in Amazonia.* Columbia University Press, New York.

Schurz, William L., O.D. Hargis, C.F. Marbut and C.B. Manifold, 1925. *Rubber Production in the Amazon Valley.* U.S. Department of Commerce, Trade Promotion Series no. 23, Washington, D.C.

Schwartzman, Stephan, 1989. "Extractive reserves: The rubber tappers' strategy for sustainable use of the Amazon rainforest." in: *Fragile Lands of Latin America: Strategies for Sustainable Development.* John O. Browder (ed.), Boulder: Westview Press, Boulder, CO., pp. 150–163.

Sen, Amartya, 1983. "Development, which way now?," *Economic Journal*, vol. 93, pp. 745–762.

Shepherd, William G., 1985. *The Economics of Industrial Organization.* Second Edition, Prentice Hall, Englewood Cliffs, NJ.

Stiglitz, J.E., 1986. "The new development economics," *World Development*, vol. 14, no. 2, pp. 257–265.

Stuckey, John A. 1983. *Vertical Integration and Joint Ventures in the Aluminum Industry.* Harvard University Press, Cambridge, MA.

Tambs, L.A., 1966. "Rubber, rivals, and Rio Branco: the contest for the Acre," *The Hispanic American Historical Review*, vol. 46, no. 3, pp. 254–273.

________. 1974. "Geopolitics of the Amazon," in: *Man in the Amazon.*, Charles Wagley (ed.), The University Presses of Florida, Gainesville, FL, pp. 45–87.

Taussig, Michael, 1987. *Shamanism, Colonialism, and the Wild Man. A Study in Terror and Healing.* The University of Chicago Press, Chicago.

Taussig, Michael, 1984. "Culture of Terror—Space of Death: Roger Casement's Putumayo report and the explanation of terror," *Comparative Studies in Society and History*, vol. 26, no. 3, pp. 467–497.

U.S. Congress, 1913. "Slavery in Peru," *House Documents*, vol. 3, 62D Congress, 3d Session, Doc. no. 1366, Government Printing Office, Washington, D.C.

U.S. Consular and Trade Reports, (various issues). Bureau of Foreign and Domestic Commerce, Department of Commerce and Labor, Government Printing Office, Washington, D.C.

U.S. Mint, *Annual Report of the Director of the Mint* (various issues, 1890–1920). Government Printing Office, Washington, DC.

Von Hassel, George, 1905. *La Industria Gomera en el Perú*. Imprenta de la Opinión Nacional, Lima.

________. 1912. "Tapping trees by electricity," *India Rubber World*, vol. 47, no. 2 (December 1), pp. 142–43.

Wagley, Charles, 1953. *Amazon Town. A Study of Man in the Tropics*. Oxford University Press, New York.

Walker, Charles, 1987. "El uso oficial de la selva en el Perú republicano," *Amazonía Peruana*, vol. 8, no. 4, pp. 61–89.

Walle, Paul, 1907. *Le Pérou Economique*. Fourth Edition, E. Guilmoto, Paris.

Walle, Paul, 1914. *Bolivia. Its People and its Resources. Its Railways, Mines and Rubber Forests*. Bernard Miall (trans.), Charles Scribner's Sons, New York.

Weinstein, Barbara, 1983a. "Capital penetration and problems of labor control in the Amazon rubber trade," *Radical History Review*, vol. 27, pp. 121–140.

Weinstein, Barbara, 1983b. *The Amazon Rubber Boom, 1850–1920*. Stanford University Press, Stanford, CA.

Weinstein, Barbara, 1985. "Persistence of caboclo culture in the Amazon: the impact of the rubber trade, 1850–1920," in: *The Amazon Caboclo: Historical and Contemporary Perspectives*. Eugene Philip Parker (Guest Editor), Studies in Third World Societies, Publication no. 32, Department of Anthropology, College of William and Mary, Williamsburg, VA., pp. 89–113.

Weinstein, Barbara, 1986. "The persistence of precapitalist relations of production in a tropical export economy: the Amazon rubber trade, 1850–1920," in: *Proletarians and Protest. The Roots of Class Formation in an Industrializing World*. Michael Hanagan and Charles Stephenson (eds.), Greenwood Press, New York, NY., pp. 55–76.

Weir, J.R., 1926. *A Pathological Survey of the Pará Rubber Tree (Hevea Brasiliensis) in the Amazon Valley*. U.S. Department of Agriculture Bulletin no. 1380, Washington, DC.

Williamson, Oliver E., 1975. *Markets and Hierarchies: Analysis and Antitrust Implications*. Free Press, New York.

Williamson, Oliver E., 1985. *The Economic Institutions of Capitalism: Firms, Markets, Relational Contracting*. Free Press, New York.

Wolf, Howard and Ralph Wolf, 1936. *Rubber: A Story of Glory and Greed.* Covici Friede Publishers, New York.

Woo, Wing Thye, Bruce Glassburner, and Anwar Nasution, 1994. *Macroeconomic Policies, Crises, and Long-Term Growth in Indonesia, 1965–1990*. World Bank, Washington, D. C.

Woodroffe, Joseph F., 1914. *The Upper Reaches of the Amazon*. New York.

Woodroffe, Joseph F. and Harold H. Smith, 1915. *The Rubber Industry of the Amazon, and How its Supremacy can be Maintained.* J. Bale, Sons and Danielsson, Ltd., London.

Yungjohann, John C., 1989. *White Gold. The Diary of a Rubber Cutter in the Amazon. 1906–1916*. Ghillean T. Prance (ed.), Synergetic Press, Oracle, AZ.

Index